The Journey through Developmental Mathematics

An Analysis of Decision Making Regarding Attitude and Time Management

Dr. Darrin Berkley

INFINITY
PUBLISHING

ISBN 978-0-7414-6910-6 Paperback
ISBN 978-0-7414-6911-3 eBook

Printed in the United States of America

Published February 2023

INFINITY PUBLISHING

Toll-free (877) BUY BOOK
Local Phone (610) 941-9999
Fax (610) 941-9959
Info@buybooksontheweb.com
www.buybooksontheweb.com

The goal of this pamphlet is to give developmental mathematics students some concrete advice regarding their "journey" through the world of mathematics. This journey is made up of the many individual decisions students make (or do not make) as they strive for success in developmental mathematics. Generally, students begin in different places on the journey, starting with the first course of developmental mathematics (Basic Mathematics) and continuing on to the last course (Intermediate Algebra). Throughout the journey there will be many occurrences of both joy and frustration. A challenge for the students is channeling their frustration.

In the math courses that I teach, students are asked about their past and present mathematical experiences. This allows me to carefully analyze the decisions that have had the greatest impact on their journey in developmental mathematics. Community college students deal with many issues impacting their academic abilities; consequently, this pamphlet will not address every issue that students face. However, this pamphlet will discuss the most common issues faced by developmental mathematics students.

The pamphlet should be read sequentially. Students should address issues in the first chapter before reading the second chapter. This may require additional communication with instructors, tutors, employers, family or school administrators.

This pamphlet will read as if you are a student sitting in my office. Since you are reading this, it is likely that you would like to do better in your mathematics courses. I'm not going to provide you with research studies in mathematics education or expect any complex tasks from you. I am going to give simple advice as well as determine the likelihood that the advice will be followed. As with any problem, it will probably require a significant amount of your time to solve. As you are going through the pamphlet, analyze the decisions that you make as well as the consequences resulting from your actions. Finally, if you are serious about succeeding at this, believe in yourself!

Let's get to work!

Chapter 1: Adjustment in Attitude

Let's try to understand your mindset. Most students that are enrolled in developmental mathematics either do not like mathematics or take no interest in mathematics. What comes to mind when you think about the course in which you are enrolled? If you don't like mathematics, try to understand why you feel that way.

One reason might be that you haven't been successful in mathematics in the past. Why not? First, do you truly believe that if you give your very best, there is a possibility that you could be successful? Don't tell yourself that you can't do it unless you've given everything you got. Also, think about those times when you *have* been successful. What did you do to succeed then?

If you believe that you can do the work, are you serious about maximizing your success? What do I mean? I am referring to the urgency of completing this journey. How important is it to you? Will your behavior illustrate the importance of pushing yourself through despite potential times of frustration? How will you deal with this? Whatever you do, don't quit!

Your success throughout this journey will likely depend on the decisions you make when you are frustrated. If you're quick-tempered, do you have a plan for controlling your temper? Speaking for all who have completed this journey, I can tell you there *will* be times of frustration. Understand that you can overcome these times. Let's try to better understand your frustration.

We need to go deeper than "I've never been good at math." We are not going to dwell on the past. Let's try to go forward with the skills that you have. As an instructor, the only concern I would have at this point is whether or not you meet the required prerequisites of the class. In other words, do you have the foundational skills necessary for success? If not, you may want to change to a lower-level developmental math course. If you have the skills but have forgotten what you have learned because you haven't taken math in several years, ask the instructor or the mathematics department if you can take an old final exam from the course prior to the one in which you are enrolled. Even though taking a prerequisite course may cost you more money, it will be beneficial in the long run for your basic foundational knowledge. Remember, you should be placed in the course that best represents the mathematical skills that you have. You will also gain more confidence by seeing that you may not be as bad at math as you think you are. Being placed in the appropriate math course will allow you to more accurately represent your ability and will likely decrease your frustration during your journey.

The next possible source of frustration may be that the explanations given by your instructor are unclear. Now you understand why having the prerequisite skills must be discussed first. If you don't have the necessary foundational knowledge, then it's possible that the explanations given by your instructor will always be unclear regardless of how many times he or she tries to explain the course content. Not understanding your instructor's explanations can be especially frustrating when you need clear explanations to complete homework or other assessments. You can solidify some of the explanations given by your instructor by asking questions or verifying that you understood the explanations.

Are you afraid to ask questions in class? Do you feel that asking questions will slow the class down? Put away your ego! Remember that you paid for the class. It's expected that you'll have questions. This is why you are enrolled in *developmental* mathematics. You need to *develop* your skills so that you're adequately prepared for college-level mathematics. You must ask questions so that you can better develop your mathematical skills. Most developmental mathematics curricula are designed so that students have time to ask questions. Anyway, your instructor manages the pacing of the course regardless of whether students ask a few questions or many questions. If too many questions are asked, the instructor may choose to answer questions during office hours or during the next class session. The instructor may also use email correspondence to answer questions. The instructor is also aware that a certain amount of content must be covered during the semester. So the challenge for the instructor is to find that balance between covering the course material and making sure that most students get their questions answered during the few hours of class time each week.

As an alternative to asking your instructor, you can use the Internet to get answers to your questions. If you don't understand a topic—for example, factoring—use a search engine such as Google or Yahoo and type in the word or phrase you're looking for. You'll likely get many links that you can use to help you answer your question. In fact, sometimes you can type in your exact question to get a response. Another option is to consider specific sites, such as Dr. Math, that might answer your questions. You may consider used books (ask your instructors—they generally have many of them) or CliffsNotes to answer your questions as well. You may try these options, but meeting with your instructors still proves beneficial.

Use your judgment. If you feel that you have too many questions to be addressed in class and you've exhausted all other possibilities, then you'll need to sacrifice some time and meet with your instructor during office hours. If the office hours given by your instructor don't fit in your schedule, then ask your instructor to meet with you at a different time. Although there are other people that you can turn to for help, it is important that you maintain good communication with your instructor. Before you meet with your instructor, make sure that you have completed all previous homework assignments in order to maximize your ability to understand your instructor's explanations. Also, make sure that you understand to the best of your ability what your instructor has already said in class. Be sure that your notes are 100% accurate. Allow the instructor to give the exact same explanation that was given to you in class. If this doesn't work, ask for an alternative explanation. Instructors often have multiple ways to explain the same idea.

What also may occur is that you might not get along with your instructor or you might simply believe that he or she doesn't have enough experience to clearly explain the course material that is causing you confusion. In that case, a new plan of action must be considered.

Is it possible to mend the relationship between you and your instructor? What exactly are the problems between you and your instructor? Remember that in order to resolve the conflict, you'll have to be clear when stating your problems to your instructor (or to a third party, if necessary). Before you speak with your instructor, understand any underlying issues that have contributed to the problems. Also, think about the decisions that you have made that may have contributed to the problems.

Understand that not all of the problems you may have are solvable. One such problem is a possible language barrier if the instructor's native language is not English. You have to decide during the first week of classes whether or not the problem is significant enough that it impacts learning.

A second problem that may not have a solution is the grading of your assessments. There is a difference between "hard" grading and "unfair" grading. If your instructor is a hard grader, he or she may have high expectations and want you to be fully prepared so that small mistakes don't carry over to other courses. Small mistakes can generally be fixed if you look over your work carefully. Instructors assign grades through the use of a grading rubric. However, some instructors don't make the grading rubrics for all assessments available to students. Feel free to ask your instructor how assessments are graded. Some rubrics do not award partial credit for wrong answers on certain assessments or individual questions. This holds true both for handwritten assessments and computer software programs.

How do you deal with an instructor who grades unfairly? First, understand that most instructors want you to succeed. (Otherwise, why teach?). Your instructor is not the enemy. An instructor typically would have no reason to "target" a student. Sometimes, the unfair grading is not always intentional if it indeed occurs. Instructors are not perfect: they will make honest mistakes in their grading from time to time. If you feel that you have been graded unfairly, do not let it go!

You must first trust that your instructor has the mathematical competence to grade your assessments fairly. If you find that you cannot have this trust, you may find yourself questioning every grade your instructor gives to you. This can create a toxic relationship between you and your instructor. In this case, it would be best if you reregister for the course with another instructor if possible.

In each of these scenarios, you must control your anger and keep your attitude in check. Strong, positive communication with your instructor is critical. Be sure that you reevaluate your assessment from both your point of view and from your instructor's point of view. Try to view your assessment through the eyes of your instructor. Was something unclear in your answer that might have caused your instructor to mark the answer incorrect? Did the instructor make a correction without actually deducting points? Understand the impact of the assessment on the overall course grade. Is it worth creating a strained relationship by saying something to your instructor simply out of anger? Don't let one incident impact the continuous communication that you should have with your instructor.

If your instructor concludes that your assessment was graded unfairly, then you should obviously receive the points back that you have lost. Do not take this personally. It is likely not the first time that an instructor has made such an error.

There are also instances when students don't realize that what they have done is incorrect. If this happens to you, ask your instructor to explain (either verbally or through the use of a grading rubric) why points were deducted on your assessment. If the explanation given for your grade remains unclear, you can ask your instructor for an alternative explanation. If this fails, your next option is to consider having another instructor—such as one who you know from a previous class—take a look at your assessment just so you can get a second opinion. If no such instructor exists, then ask the department chair to give you an explanation for the grade you received. You can try to make a case for why you think you were graded unfairly, but do not be upset if the situation is not resolved in your favor.

In summary, be sure that you clearly understand the grading policies of your instructor. However, your main focus must be on *learning* the course content. An instructor can often tell which students genuinely want to learn and understand what they are being taught and which ones are just trying to get enough points to get a passing grade. If you have no interest in learning, then it will be difficult to succeed in any course, not just a developmental mathematics course. The challenge in having "an attitude to learn" is finding that math topic that sparks your curiosity. This represents another source of frustration for students: working hard to learn something that will never be used outside of the classroom. If a student doesn't see the usefulness of a particular topic, then the topic will likely be forgotten, which will create problems in future mathematics courses. So it is important for instructors to emphasize the importance of gaining a strong foundation in mathematics to alleviate possible frustration in the future.

It is probably true that if your major is art, you probably will have no use for the quadratic formula outside of the mathematics classroom. A future nurse doesn't recognize the importance of factoring in everyday life. When would I need to combine like terms as a business major? These are questions and statements that have come from developmental math students. At first glance, it's hard to see why these topics would be relevant to real life. Let's think a bit about why you are being taught mathematics.

I agree that at times you will be taught mathematics topics that you will never use. Remember, I am being honest as a mathematics instructor. But I could also make the case that very few courses are made up of topics that *all* students can use outside of the classroom. For example, do I need to know what a verb is to be a productive citizen? Do I need to know the periodic table of the elements? Do I need to know anything about the American Civil War in 1861? How will any of this help me now?

In practically all disciplines, we will always be taught more than we actually use. But we should all agree that it is better to have too much knowledge than not enough knowledge. Let's not look at mathematics as a set of discrete, disconnected topics. Instead consider mathematics as a language that all of us must use at some point in our lives. In thinking of language as a method of communication, you can probably think of many words that you would never use in real life. Understand that the students who are going through the developmental mathematics journey with you—those who sit next to you in class—come from all walks of life. Some want to be teachers, others want to be mathematicians, and some don't know what they want to be. The mathematics curriculum has to accommodate all of these students. This is by no means an easy task. The mathematics curriculum is not intended for any particular group. It is intended to develop basic knowledge that you are expected to know. Let's think about the skills that you strengthen other than mathematical ability. Why do you even need to take mathematics at all?

Some of the many uses of mathematics that are relevant to all students are the understanding and study of quantity, change, reasoning, structure, measurement, patterns, and risk. If you look at these topics carefully, you'll discover that each of these skills are used in the lives of all productive citizens. Many of these skills are learned indirectly and are absolutely essential to understand. This is why mathematics is taught at every level of schooling and why it must be understood on the college level in order to receive a college degree.

What else do you learn in the developmental mathematics classroom? It is important that you believe that skills can be learned during the journey in developmental mathematics. Since mathematics is often viewed as a "black-and-white" discipline, there is little room for error. Again, this creates frustration for students. You probably have experienced times when you have spent hours working on one problem only to find out that it was wrong due to a very small error. You may also have spent long hours completing an assignment only to key the answers into the computer mathematics software incorrectly and receive no credit. This kind of incident is more common in mathematics and science than in other disciplines. Of course, students generally don't question the importance of accuracy unless it has an impact on their grade. Whether in school or in the workplace, when tasks are to be completed, one expects the task to be completed accurately. You may get frustrated at an instructor when he or she takes of points for inaccuracy, but consider an inaccuracy in your paycheck—to your disadvantage—and think about the importance of accuracy.

Another important skill you learn is how to take a rule and then apply it to unknown situations. This is a skill that you probably already have. Mathematics can be used to strengthen this skill. You can think of rules as those that have been given to you by parents, those that are decreed by law, those given by your employer, and many more. These are rules that you may not understand completely, but you try your best to apply them. You are given many rules in developmental mathematics, and you often have to apply multiple rules to a single problem. Not understanding one rule could have a significant impact on the entire problem. Simply trying to memorize all of the rules will not allow you to apply the rules in various contexts. This is an idea that is prevalent in problem solving. Word problems often serve as those "unknown" situations. Understanding how to apply rules can be a useful skill in real life as well. One of the most common expectations for most jobs is likely to be an ability to solve problems.

Mathematics can also improve your critical thinking ability. Critical thinking has many definitions, but some of the characteristics include determining relevant and irrelevant information, making accurate judgments based on the problem context, questioning assumptions, and using inductive and deductive reasoning. All of these characteristics are used outside of mathematics as well. Some argue that critical thinking is not used in developmental mathematics as often as it should, but there is no doubt that critical thinking plays a significant role in education in general as well as in mathematics.

What does mathematics offer that may be relevant to you? One skill that you strengthen is your patience. How much patience does it take to continue taking mathematics courses that do not count for credit? If you are enrolled in a developmental mathematics course, it is likely that your patience has been tested on numerous occasions. Some students find themselves losing confidence in their ability and losing patience as a result of poor results on tests or other assessments. The question arises as to what decisions students should make to address this issue. It is hard to convince students to remain patient when they continue to receive poor results over long periods of time. How much do you give before you just quit? This is a question that you may have asked when completing homework assignments. Some students quit if they can't solve a problem on the first attempt. Others make numerous tries before quitting. Ask yourself how you feel when you are in these situations. Do you try to get assistance when you struggle? Or do you start to tell yourself that you just can't do it?

How have you handled situations similar to this outside of the mathematics classroom? Did anyone or anything contribute during the times when you remained patient? Another good idea would be to read about other people who had to remain patient to achieve a goal. It can take a lot of hard work, but you can do it if you apply yourself and think about the end result. Remember that you are not by yourself. Whether they have an educational background or not, family and friends can play a supportive role in helping you maintain your patience.

Some students feel that if they can succeed in mathematics, then they can succeed in anything. Learning mathematics is a challenging task, but I'm sure you know that understanding mathematics is not impossible. Mathematics may not come to you as quickly as it comes to others, but it is not a race. Be patient and understand that with your failures can come great success. That success is having enough patience to complete the journey.

We have just discussed some of the reasons why mathematics should be important to you. Mathematics is all around you! Since most companies use formulas to some extent, understanding those formulas does not just affect those that devise the formulas—it will likely impact you as well since your life can be affected by the formulas. So do not have the attitude of "Somebody else needs to understand those formulas." Counting on someone else's knowledge when you could obtain that knowledge yourself can cost you thousands of dollars in the future. Many community college students change majors. You never know when you'll change your major to one in which mathematics will play a greater role.. Don't take your chances! Learn now when you have the opportunity. It may not present itself again.

So what if you choose not to learn mathematics? What are the consequences?

Well, one of the obvious consequences is that you will not get the degree that you worked hard to obtain. On many occasions, students put a great deal of time (not necessarily productive) in their mathematics courses. An associate degree (or any degree for that matter) requires a lot of dedication and drive. It will require you to do things that you do not want to do. Because of the importance of mathematics, it is a requirement for your degree. Colleges and universities expect students to be literate in mathematics and science to compete in a global economy.

Second, it is just a waste of valuable time. All of us believe that our time is valuable. If you're going to take a developmental mathematics course, why not give it your best the first time? There could be many answers to this question. In many cases, the student knows what needs to be done to succeed. The student must have the mindset to learn the mathematics content even if he or she does not have an interest in mathematics. Sometimes, students try to memorize the course material because it is easier to accomplish and saves time. The problem with that is that there is often too much material taught in a semester to memorize all of it. Also, just because you have memorized how to do a problem doesn't necessarily mean that you understand it. Memorization can get you through some courses, but it will eventually catch up to you if you choose not to learn. Memorization will save you time in the short term, but will likely result in a major loss of time in the long term.

Third, it is a waste of money. Even if you are on a needs-based scholarship or grant, someone's money is still wasted. It has an even bigger impact if you pay for your own education. Most students will admit that they don't have money that they can just waste or throw away. If you're getting help from parents, your parents may decide that you are not serious about your education and choose not to pay for your classes anymore, causing you to drop out of college.

Fourth, you may not get accepted into certain academic programs. Some programs require that you complete a certain number of mathematics courses with a minimum grade for acceptance. Remember that you are competing with many other students. Generally, the hardest-working and most knowledgeable students are likely to get into such programs.

Fifth, a failing grade can take a huge hit on your GPA. Often a certain GPA is required for acceptance into certain programs. Also, some employers will look at your GPA for employment purposes. At some point, you may consider applying to graduate school. Admissions departments consider GPA as a factor for acceptance into graduate programs.

Finally, what does a failing grade tell you about yourself? What does it tell others about you? Often it represents failure from different perspectives. For those who really don't understand what the "F" (or "D") means on your transcript, you still may have to answer to parents or other family members. Students who live with their parents may have to answer to them, which may not be a pleasant experience. Older students who fail may have to answer to their spouses because of the economic impact on the family. Sometimes, a degree may be a new requirement for a current job. Not having a degree may result in the loss of future opportunities, which could have serious life consequences. One way to avoid these consequences is to simply be a good student. In other words, just do what your mathematics instructor asks you to do.

As an instructor, I'm going to share some things you can do to prove that you are being a good student. If you are willing to be a good student, I expect you'll have success in developmental mathematics. Ask yourself if you are willing to commit to these tasks. If so, promise yourself that you will make the necessary sacrifices to complete the following tasks. Make no excuses for not completing the tasks. This will show that you are serious about completing your journey.

1. **Focus in Class**

   Stay awake; cell phones should be put away during class; come to class prepared to learn; keep side conversations to a minimum

2. **Come Prepared for Class**

   Textbook; several pencils (erasers); required calculator; notebook

3. **Adjust to the Pace of the Class**

   Student must make adjustments

4. **Allot Time to Do the Work**

   At least **2 hours** per day (14 hours per week) outside of class is necessary. This is essentially a part-time job.

5. **Handle Frustration Appropriately**

   Get the help you need. Speak with your instructor regarding **any** concerns. Use the math lab. Learn from your mistakes.

6. **Ask Questions**

   Put away your ego. If you knew it all, you wouldn't be here. Don't worry about what others think. If you don't know something, tomorrow may be too late to ask.

7. **Handle Transportation Issues**

   Leave for class at least 10 minutes to allow time for traffic and parking. Plan on staying for the entire class.

8. **Complete Work in a Timely Fashion**

   Respect the due dates. Late work is not accepted. "I forgot" is unacceptable.

9. **Understand Errors Made on All Assessments.**

   Students are held to the highest standards regardless of ability level.

10. **Understand the Consequences of Your Actions**

    Whether the results are good or bad, hold yourself accountable!

11. **Read All Assigned Sections in the Textbook**

    These present the main ideas in the unit and cover the topics that will most likely be on your exam.

12. **Complete a Practice Exam**

    If you have math anxiety, it is important that you simulate the actual exam and analyze any errors.

Chapter Summary:

This chapter discussed the importance of having an attitude of learning. You must take the course seriously, and it should show in your actions and decisions throughout the journey. During this journey there will be times of frustration. How you handle your frustration will impact the journey. Some of the frustrations include previous negative results on mathematics assessments, unclear explanations by your instructor, language barriers between instructor and student, poor student-instructor relationship, and grading concerns. Another frustration that was addressed was the relevance of mathematics. In other words, why do you even need to take a mathematics course? Once the relevance of mathematics is recognized, then students must understand the consequences of not understanding mathematics. These consequences are not pleasant to face. You can avoid these negative consequences by being a good student and doing what you are asked to do. The next chapter assumes that your attitude of learning has been established, and we will move on to discuss the impact of time management.

Chapter 2: Time Management

Now that you have made any necessary adjustment in your attitude toward developmental mathematics, we'll look at the amount of time you can devote to your coursework. The first step is to create a schedule for every day of the week including everything that occupies any of your time. Do not forget to include transportation time to work or school. As an estimate, look at your schedule to determine if you have a minimum of at least 14 hours per week for your mathematics course. Why 14 hours? Remember that you will have to consider the amount of time it takes for completing homework assignments (this includes some challenges that might occur in solving problems), reading a section of a textbook multiple times prior to class to maximize understanding, studying for examinations, completing of small projects, reviewing your notes, understanding computer software, attending tutoring sessions and office hours, reviewing previous graded assessments, and so on. Note that 14 hours per week is similar to a part-time job, but you may or may not need all those hours. It is better to overestimate the number of hours you need for your developmental mathematics course than to underestimate. This estimate will change depending on your mathematical ability. Also, if you have a tutor working with you from the first day of the course, this may minimize your time. Again, to be safe, I would be hesitant to enroll in a developmental course if you can't devote 14 hours per week to the class. If the mathematical topics are more challenging than you anticipate, if you have a poor student-instructor relationship, or if it has been a long time since you have taken a mathematics course, expect to devote at least 2 hours per day or more. Are you willing to put in this amount of time to pass your developmental mathematics course?

To get a better estimate of your time, you'll need to analyze both speed and accuracy when completing assignments. Time yourself when given an assignment. How long does it take for you to complete an assignment? What percentage of questions did you get correct? Generally, you are given the answers to your homework assignments so you can confirm that you have completed your homework assignments correctly. If you feel that it is taking you too long to complete assignments, then speak with your instructor to verify that you have a thorough understanding of the intermediate steps involved in solving problems. Have your instructor give you a problem and allow your instructor to watch you as you complete the problem. Remember that checking your work will decrease speed but increase accuracy. It is strongly recommended that you take the extra time to check answers when possible. This will allow you to minimize the number of careless mistakes that you might make.

Suppose you don't have this amount of time to devote to your mathematics course. What is your next course of action? Because time is so important, I would recommend that you reregister for the course when you have this amount of time to devote to the course.

One activity that often takes a significant amount of time is your work schedule. This is a tricky topic to address. As a community college student, most students must work. Often there is no other option because many students pay for their education out of their own pocket. It's hard to tell students to consider working fewer hours because often they need the money to support their family. The question often arises, what is more important, work or school? Sometimes, one or the other has to be given less time, which can result in ineffectiveness. Many students consider work more important because without work they can no longer pay for school. With the impact of the bad economy, it is hard to turn down opportunities to work more hours. So some students decide to take the risk of working long hours and devoting little time to their mathematics courses. If it is possible, working fewer hours can give students additional time to complete their mathematics assignments. Other factors that may affect the decision to work fewer hours are whether or not homework can be completed on the job, the flexibility of the work schedule, and the student's overall financial situation.

Students may also allot more time to their mathematics course by enrolling in fewer courses. Many students underestimate the amount of time required for particular courses. As with your mathematics course, speak with the instructors of the other courses in which you are enrolled before the courses begin. Also speak with some students who have already taken these courses. This will give you a better idea as to how much time you'll need for the courses. This will also help you revise your schedule so that you know if you have enough time to devote to all your courses.

A challenge that many students face is a rush to graduate. Students aren't always patient or in some cases they just really want to get into certain programs at a particular time. There is an urgency to complete courses as fast as possible, especially when students are enrolled in school part-time. Sometimes students get an increase in pay once they get a degree. How do you convince a student that he or she is enrolled in too many classes? Students do get recommendations from the advising department, but some students do not necessarily take the advice of the advisors. Ask yourself if there should be a rush to graduate. What is your ultimate goal when you graduate? What opportunities does graduation give you? Will those opportunities always be there or must you move quickly to graduate? Answers to these questions will vary according to the individual. There will be a certain amount of risk that each student must consider when making the decision to take a significant number of credits.

The amount of time you have for your mathematics courses may also be affected by transportation. If you don't have a reliable means of transportation, you will likely be at a huge disadvantage. Not all places have convenient public transportation, and using a taxi to get back and forth to classes may not be affordable. Even though you could rely on others, remember that it is *your* education. If you are late or absent for a mathematics class, it can be very hard to recover. It will be you who suffers because of someone else. If you don't have a reliable vehicle, what can you do to get a vehicle (even if it is a used vehicle)? Can you buy a used (working) vehicle from a dealership or auction? Can you save some money or borrow some money to purchase a vehicle? If this is not feasible, can you use a family member's vehicle to travel to and from classes? Do you know someone who travels to school and leaves school at the same time that you do?

How does this impact time in your mathematics course? You may need to see your instructor or tutor outside of class time, which means that you might have to travel back to school for the appointment. If you don't own a computer and you're enrolled in a mathematics course that requires completion of online homework, you may need to spend several hours in the library—and you cannot guarantee what time you'll be finished. You may need to take advantage of study groups to prepare for your mathematics tests. Again, this would be hard to accomplish without a reliable source of transportation. Study groups can be beneficial in reducing the amount of time it takes to complete assessments or prepare for an exam.

Your family also can impact the amount of time that you can devote to your mathematics course. This can include caring for sick relatives, taking care of children, sharing a single vehicle with relatives, doing chores or tasks required by family members, and so on. If your family currently requires a lot of time, it may not be the best time for you to enroll in community college. Only you know the amount of time that your family members require, so I'll only bring to your attention the importance of your family. You may need to have a discussion with your family members about the importance of your schooling if you indeed want to be successful in your courses. They might allow you more time for your mathematics course and your other courses. Do you have babysitting in place for your children if needed? Do your children or any of your relatives have special needs that require a great deal of your time?

Finally, there could be a variety of personal issues that this pamphlet cannot address. If you have special needs or have to make frequent visits to the doctor, this all must be considered before considering taking a mathematics course. Each student has to determine what is most important in his or her life. The significance of these personal issues on a student's education can vary greatly. Personal issues can also include any addictions that could impact a student's life. What plan do you have to address these issues? If you can't avoid these issues, think of how you could minimize them.

A lot of factors have been discussed here that can impact your time. What is most important to you? What takes up most of your time? You will have a hierarchy of tasks. Determine the order of importance of these tasks. Where does school fit in this hierarchy? More specifically, where does learning (not memorizing) the topics in your developmental mathematics course fit into this hierarchy? Since most students who read this do not have an interest in mathematics, it may go lower in the hierarchy than other things. If this is true, it is also possible that it won't get enough of your time.

After a review of what takes up your time, the next question to ask is if you are willing to devote the time to your coursework. Some students don't have any of the issues discussed above. They simply say, "No, I am not going to do the work because I don't want to do it." Do you believe that you're just lazy when it comes to completing work in mathematics? These students would rather give their time to other tasks that they enjoy. They have never liked math and never will. They may also find that they don't want to work hard at something in which they think they won't succeed. The decisions that students are asked to make in this pamphlet *will* require some sacrifice. Think about the hierarchy you considered above and ask yourself why you created it in such a way. As with work, school is not always going to be fun. Think of the many things that you must do that are not fun. Hopefully, you are not deceived by reality TV. Their reality is not going to be your reality. Time has to be given for hard work that is not always immediately rewarding. Try to look forward to the potential benefits of completing the journey.

Chapter Summary:

Students need time to succeed. Students can do well in their courses if they have enough time to complete all of the assignments. A very detailed schedule is essential for students to truly determine if they have enough time to give to their mathematics course. Tasks that have the greatest impact on the amount of time students have for their mathematics courses are number of hours worked per week, number of courses enrolled in along with the developmental mathematics course, transportation, family concerns, and other personal issues. Students must decide where completion of a developmental mathematics course fits in their many priorities. Students must also decide if they are willing to devote the time to the course if it fits in their schedules.

Chapter 3: Conclusion

This pamphlet is intended to help developmental mathematics students answer many questions and make many decisions throughout their journey through developmental mathematics. This pamphlet by no means covers every possible question you might have, but the questions addressed here are very common based on my experience reading hundreds of responses about student issues during this journey.

It should also be noted that there are no recommendations here about academic ability or any specific developmental course or topic. There are many books that already address these issues. This pamphlet intends to address two significant nonacademic issues that have an impact on student success in developmental mathematics.

These questions should allow students to consider whether or not the time is right to enroll in developmental mathematics. Not all students are prepared for college, or, more specifically, for developmental mathematics. An attitude of willingness to learn needs to be in place prior to enrolling in developmental mathematics. As with other tasks, mathematics takes time. Students have to be sure they have the time to give before enrolling.

Following the advice in this pamphlet does not guarantee an A in a mathematics course, but it is very likely that a student who follows these recommendations will receive a passing grade. This is not to say that your goal in any course should not be to work hard for an "A" grade. In fact, an adjustment in attitude should tell you that you can do the work if you put your mind to it and remain focused. To add to this advice, understanding the attitude and time management strategies of other "A" students in developmental mathematics is also important. In conclusion, continue talking positive and believing that developmental mathematics is something that you can conquer at this point in your life. Make no more excuses, fight through the obstacles, and complete the journey!

Made in the USA
Monee, IL
13 September 2026

62977542R00020

Building Consulting Skills for Sport and Performance Psychology

Building Consulting Skills for Sport and Performance Psychology is an international compilation of case studies that introduces students and early career professionals to pertinent issues within a variety of performance environments.

With contributions from prominent international practitioners, the book offers case studies from a variety of performance environments including sport, performing arts, and high-risk occupations. Each chapter presents a case study and provides exploratory questions relating to ethical considerations, intervention planning, and awareness of diverse themes of race, culture, gender, immigration status, and disability, among others. These reflection questions encourage students to acknowledge the individual circumstances of all performers and to recognize their own unique background and how this impacts their attitudes, beliefs, and connection to potential clients. The text promotes honest self-reflection regarding consulting strengths and potential areas for improvement, and an online Instructors' Manual looks at possible solutions for each case.

The book will be essential reading for supervisors, mentors, students, and professionals engaged in providing consulting services in distinct performance environments. It encourages its readers to use careful self-reflection to integrate theory and practice and refine their own consulting philosophies.

Sarah L. Castillo, a CMPC on the Registry of Approved Mentors, has focused her career in both academic and applied practice settings. She earned her PhD from the University of Idaho in 2000, and has served on the executive board for the Association for Applied Sport Psychology. She currently serves as the Director of Sport and Performance Psychology programs at University of Western States, USA.

Chelsea Butters Wooding is an Assistant Professor of Exercise Science at North Park University in Chicago, IL, USA. She is a CMPC through AASP and holds a PhD in Sport and Exercise Psychology from West Virginia University, along with an MA in Counseling from WVU, an MS in Kinesiology from California State University, Fullerton, and a BA in Psychology from California State University, Long Beach.

Douglas A. Barba is an Associate Professor in the Psychology Department at National University, USA, where he serves as the Program Director for the BA in Sport Psychology and MA in Performance Psychology programs. Doug earned his PhD from the University of Florida and an MA in Psychology from the School of Human Behavior at United States International University in San Diego.

Stiliani "Ani" Chroni earned her PhD from the University of Virginia in 1997 and pursued a dual career as an Academic and Applied Sport Psychology Practitioner. Currently she is a Professor in Sport Psychology and Sport Coaching and the Director of the Sport and Social Sciences Research Group at Inland Norway University of Applied Sciences, Norway.

Building Consulting Skills for Sport and Performance Psychology

An International Case Study Collection

Edited by
Sarah L. Castillo,
Chelsea Butters Wooding,
Douglas A. Barba and
Stiliani "Ani" Chroni

Routledge
Taylor & Francis Group
NEW YORK AND LONDON

Cover image: Getty

First published 2023
by Routledge
605 Third Avenue, New York, NY 10158

and by Routledge
4 Park Square, Milton Park, Abingdon, Oxon, OX14 4RN

Routledge is an imprint of the Taylor & Francis Group, an informa business

ISBN: 9780367545420 (hbk)
ISBN: 9780367545406 (pbk)
ISBN: 9781003089629 (ebk)

DOI: 10.4324/9781003089629

Typeset in Bembo
by codeMantra

Access the Support Material: www.routledge.com/9780367545420

Contents

Contributors

Frank Eirik Abrahamsen earned his Ph.D. in 2007 from the Norwegian School of Sports Sciences. For 10 years, he worked full-time with the Norwegian Olympic Training Center along with his part-time academic work. Now, he works full-time as a sport psychology researcher while he still does a lot of applied performance psychology work, mainly in sports but also with performers in music, TV, and military operations.

Kaori Araki, Ph.D., earned her Ph.D. from the University of North Carolina at Greensboro in 2004. She is a retired professor and a chief consultant at CORAZON Co., Ltd. in Kyoto, Japan. She is vice-president of the Asian South-Pacific Association of Sport Psychology. She has consulted with the Japan Rugby national team and contributed to the historical wins at Rugby World Cup 2015 in England. She is currently consulting with Handball, Hockey, Rugby, Track and Field, and Volleyball teams as well as athletes, singers, and business persons.

Elizabeth Beaubrook is a Mental Performance Coach with nearly a decade of experience working with both elite collegiate and tactical athletes. She earned her Master's degree from Miami University and is a certified mindfulness trainer. Liz has trained over 3000 military personnel in the US Army and Air Force, and currently serves as a Cognitive Performance Specialist training Air Force Pilots in San Antonio, TX.

Sebastian Brueckner earned his Ph.D. at the University of Tennessee in 2006, and has established himself as an internationally recognized expert in applied sport psychology. Currently, he is working in private practice as a mental performance consultant and coach, and provides consultancy to ongoing projects funded by the German Federal Institute of Sport Science (BISp) and the European Union.

Koen de Brandt is an assistant professor at the Vrije Universiteit Brussel (VUB) where he earned his Ph.D. in Sport Science in 2017. His research focuses on athletes' holistic development and dual career in sport. Since 2013, Koen has coordinated the university department

'Top Sport and Study' of the VUB. Before that, he was active as a professional volleyball player.

Megan Byrd, CMPC, is an Assistant Professor in the Health Sciences and Kinesiology Department at Georgia Southern University, a member of the University's Mental Performance Team, and the Co-Director of Mental Performance for South Georgia Tormenta FC. Megan has consulting experience with a variety of sports and athletes from youth to professional levels and business settings. In addition to consulting, she supervises graduate students in training and teaches undergraduate and graduate sport psychology courses.

Sarah Clifford is a Mental Performance Coach and owner/founder of Performance Mind Consulting LLC. She has personal experience in the elite athlete world, as well as multiple years working with performers of different backgrounds, ages, and abilities. Sarah is particularly passionate about teaching skills that athletes can use not only to excel in their current endeavors, but in life outside of sport as well.

Ashley Coker-Cranney, LPC, CMPC, is the Executive Director of Whole Brain Performance in Morgantown, WV where she splits her time between clinical practice and clinician training, specializing in helping clients redefine their relationships with success, failure, fear, change, and stress to enhance their performance. As a mentor, Dr. Cranney directs the internship training program and provides CMPC consultation to mental health clinicians and counseling programs. Dr. Cranney's advanced training in EMDR, ACT, and somatic psychology supports her service to clients' and trainees' needs across their whole systems so that they can utilize flexible frameworks at each level of processing to live wholly, congruently, and freely in pursuit of what matters to them.

Simon Defruyt is a Sport Psychologist working at the Vrije Universiteit Brussel where he graduated as Doctor in Psychology in 2019. At the VUB, he combines his research on the topic of dual career with valorization initiatives (e.g., dual-career consultancy, education for dual-career support providers, and follow-up on online tool DualCareerTools).

Poppy DesClouds is a Mental Performance Consultant with the Canadian Centre for Mental Health and Sport and a member of the managing council of the Canadian Sport Psychology Association. She earned her Ph.D. in the psychosocial sciences of sport and her MHK (Intervention and Consultation) in the School of Human Kinetics at the University of Ottawa. Her research interests include applied sport psychology, mental health, and the intersection of technology and well-being.

Natalie Durand-Bush is a Sport Psychology Professor in the School of Human Kinetics at the University of Ottawa in Ottawa, Canada. She specializes in mental skills training, mental health, and coaching psychology. Natalie co-founded and chaired the Canadian Sport Psychology Association and served as President of the Association for Applied Sport Psychology. She has been working as a Mental Performance Consultant for the past 27 years, and she is currently the Executive Director of the Canadian Centre for Mental Health and Sport.

Trevor J. Egli is a Professor and Program Director of Sport & Fitness Leadership at Johnson University-Tennessee, and he also does applied work with athletes and teams. He earned his Ph.D. in Sport Psychology and Motor Behavior from the University of Tennessee–Knoxville and is a current CMPC. He and his wife, Meagan, have two young children, Beau and Isla.

Dr. Andrea Faull-Brown, a Chartered member of the BPS and BASES, has a breadth of knowledge and experience in both research and applied practice, spanning two decades. She earned her Ph.D. from Cardiff Metropolitan University in 2010, and has she supported athletes and coaches in their preparations for the highest level of performance, including athletes at multiple Paralympic Games. She currently serves as the Head of Sport and Exercise Science at the University of Worcester, England, UK.

Kathy Feinstein is a Licensed Mental Health Counselor and Certified Mental Performance Consultant in private practice in Naples, Florida. Her practice focuses on athlete mental health and performance. Kathy earned her Master's degree in Psychology from Nova Southeastern University, and she is currently a doctoral candidate in Counselor Education and Supervision at Regent University.

Leslee A. Fisher is a Professor Emerita and prior Director of the Sport Psychology and Motor Behavior Graduate Program at the University of Tennessee. She holds a Ph.D. in Sport Psychology (UC Berkeley), an MEd in Counselor Education (University of Virginia), an MS in Education (University of Michigan), and a BS in Education (University of Michigan). Her research focuses on feminist cultural sport psychology and caring coaching practices. Leslee is also a Fellow in the Association for Applied Sport Psychology (AASP), an AASP-approved mentor, served as AASP's Secretary/Treasurer, and is a Certified Mental Performance Consultant (CMPC).

Doug Gardner has been a pioneer in the field of Performance Psychology for over 24 years, consulting with individuals in a variety of high-performance environments, from youth and professional sport to corporate leaders and organizations who seek to create an empowered culture of success through individual and collective development. From his work with the Boston Red Sox, the NFL Players Association, and with competitive youth sport athletes seeking Olympic, collegiate, and professional careers, Dr. Gardner continues to venture into new high-performance ecosystems, such as professional esports where in 2018 he was hired by Immortals to create the first player development department in professional esports, addressing the personal and professional growth of their players and coaches.

Henrik Gustafsson is a Professor in Sport Sciences at Karlstad University (Sweden) and the Norwegian School of Sport Science (Oslo, Norway). He also works as a sport psychology consultant with elite athletes in both individual and team sports.

Brandonn Harris, CMPC, LPC, NCC, is the Program Director and Professor of Sport & Exercise Psychology at Georgia Southern University. He has extensive experience providing mental skills training as well as clinical mental health services with a variety of populations ranging from youth to professional athletes and teams.

Jacob Jensen is an Associate Professor in the Kinesiology Department at California State University, Northridge. He teaches courses in sport psychology and supervises graduate students working toward their CMPC certification. Jacob earned his Ph.D. in Sport Psychology from the University of Tennessee and an MS in Sport Psychology from the University of Utah.

Abby Keenan, MS, CMPC, is a mental performance coach and entrepreneur (Intrepid Performance Consulting, LLC) in the Atlanta, GA and an adjunct faculty member and CMPC mentor for the Sport Psychology Department within the JFK School of Psychology at National University. She has over a decade of consulting experience, working with athletes aged 11 and up and supporting US Army Special Operations Soldiers at Fort Bragg, NC. She specializes in supporting teenage student-athletes, helping them to feel seen, heard, and understood, learn about themselves, build confidence, perform better, and ultimately, thrive.

Göran Kenttä earned his Ph.D. in Psychology at Stockholm University in 2001. Currently, he holds a research position in Stockholm with the Swedish School of Sport and Health Sciences, and an applied practitioner position with the Swedish Sport Federation as the Head of Discipline in sport psychology.

Matt Lacoste is a psychologist in private practice in France, Canada, and the USA. He specializes in sport psychology and eating disorders. Dr. Matt earned his Ph.D. in France at the University of Toulouse, and he has been teaching sport psychology for 10 years.

Alexander T. Latinjak holds a Ph.D. in Psychology from the Universitat Autònoma de Barcelona. His main activities are academic in nature, including teaching and researching applied sport and exercise psychology at the University of Suffolk (UK). However, he continues to practice as a sport psychology consultant working with individual athletes, coaches, teams, and sports club management.

Karen Lo is a CMPC® on the Registry of Approved Mentors and the first CMPC® in the greater China area. She previously served on the AASP International Relations Committee and the CMPC® Marketing Task Force. A Hong Kong native, she spends her time working in elite and youth sport settings and provides psychological support to world-class jockeys and youth golfers at the Hong Kong Jockey Club and Hong Kong Golf Association. She is currently a part-time lecturer at the Education University of Hong Kong where she teaches sport psychology at the undergraduate and graduate levels. Her most recent Chinese publication, *Sport Psychology: Building Confidence and Maximizing Potential*, is the first sport psychology book in Hong Kong and was shortlisted for the 10th Best Books book award by the Hong Kong Teacher-Librarians' Association.

Carolina Lundqvist earned her Ph.D. at Stockholm University in 2006. She is an Associate Professor in both Psychology and Sports Sciences and a Licensed Clinical Psychotherapist. Her current academic appointment is with the Department of Behavioral Sciences, Linköping University where she is also part of the research group Athletics Research Center. Carolina has extensive experience in applied sports psychology work as well as clinical psychology with elite sports and has participated at several Olympic Games as a sports psychologist.

Charlie Maher is the Sport and Performance Psychologist for the Cleveland Guardians Baseball Organization, and Professor Emeritus of Psychology at Rutgers University. A licensed psychologist and Certified Mental Performance Consultant (CMPC), he has 33 years of experience in providing sport psychology program services to a range of professional sport organizations, collegiate athletic departments, and elite sport academies worldwide.

Thierry R. F. Middleton earned his Ph.D. from Laurentian University in 2021 while funded as a Joseph-Armand Bombardier Scholar. He was the 2019 recipient of the Association for Applied Sport Psychology Student Diversity Award, a 2021 Governor General Gold Medal recipient, and was the inaugural Junior Editor-in-Chief of the *Journal for Advancing Sport Psychology in Research*. He is currently a Lecturer in Sport and Exercise Psychology at the University of Portsmouth in the United Kingdom.

Adam H. Naylor is the Leader of Performance Psychology for Deloitte Services LP. He earned his Ed.D. in Counseling Psychology and M.Ed.

in Counseling with a specialty in Sport Psychology from Boston University.

Doug Newburg was Faculty at the University of Virginia School of Medicine, USA, for 15 years, where he was named to the Academy of Distinguished Medical Educators. He has also interviewed and worked with hundreds of performers and professionals in fields as diverse as sports, health, business, the military, medicine, and education.

Jen Schumacher, Ph.D., CMPC, is the Mental Performance Consultant for Northwestern University Football. Jen is a professional speaker, and she provides coaching and mental performance training for a variety of elite performers, including professional athletes, Fortune 100 executives, NCAA Division 1 teams, cadets at the United States Military Academy at West Point, and Special Forces Officers in the US military. A performer herself, Jen recently swam the English Channel, and she is a CrossFit athlete.

Damien Stewart is a former elite athlete who is now a Sport & Exercise Psychologist in private practice based in Queensland, Australia. Prior to becoming a psychologist, Damien was a Police Officer at both the State and Federal levels for 20 years and now works with Police Officers as a psychologist in the mental health, and particularly trauma, space. Damien is also a Senior Lecturer in psychology at the Institute for Social Neuroscience in Melbourne and is completing a Ph.D. on the Transmission of Intergenerational Trauma at LaTrobe University.

Grzegorz Więcław holds an MSc and is certified as a Specialist in Applied Sport Psychology (SASP-FEPSAC), and graduated from Simon Fraser University (Canada) and University of Jyväskylä (Finland). He is a self-employed applied sport psychology consultant based in Gliwice, Poland. In his professional practice, he serves elite and youth elite sports clientele representing a wide range of sport disciplines.

Paul Wylleman is a Full Professor in Sport Psychology at the Vrije Universiteit Brussel where he graduated as Doctor in Psychology in 1996. His research focuses on holistic development and dual careers in elite sport. For more than 30 years, Paul has provided psychological support to athletes, teams, and coaches and was team psychologist to the Dutch Olympic TeamNL at the 2016 Rio and 2020 Tokyo Games.

Shameema Yousuf is a UK Practitioner Sport Psychologist and Mental Health Therapist in private practice, a published author, and scholar in performance, wellbeing, and cultural sport psychology. She works across Olympic, professional, and youth sport internationally, and shapes culture and performance in the corporate industry. Shameema earned two Master's degrees with an MSc from the School of Management, Brunel University, UK, and an MEd from the Wheelock College of Education and Human Development, Boston University.

Preface: Bridging Theory with Action: Voices and Perspectives from Around the World

Stiliani "Ani" Chroni, Chelsea Butters Wooding, Douglas A. Barba, and Sarah Castillo

Following data collection for my doctoral research, a polo player I had interviewed and whose polo rating had been lowered a month earlier became interested in developing the mental aspect of his game and asked me to work with him. So here I was, once again, in sunny West Palm Beach, literally feeling happy-go-lucky with a tad of nervousness as this was my first "respectfully" paying client, watching a 22-goal team practice, and paying particular attention to one poloist's game. As Adam approached the sidelines to change horses, he also came up to me and said, "So, what do you see?" In that moment I missed a couple breaths (which happened to me once again, on a different occasion, while working with poloists)! Twenty-five years later, I can still vividly remember the time of day, the sun, the field, the players, and where exactly I was standing asking myself, "Why isn't Bob Rotella here with me?" I regained my breath, told myself "You need to answer," and I did. I was seeing hesitation when he was up with the ball, it was noticeable in the few extra seconds he used to scan the field, his teammates and opponents, in his body posture when shooting at goal or making a pass. We agreed to talk more after practice, and we did. Later in the day, I drove to Boca Raton where the nearest Barnes & Noble bookstore was located to get a sport psych book. Since Doc wasn't there with me, I thought a book would be the next best thing. The closest book I found was about putting; I was in Florida after all, the golf state! *The Mental Art of Putting* (1995) was written by Patrick Cohn and Bob Winters. Bob Winters was still around UVA when I started my PhD – his nick name was "Pretty Boy" for his always perfectly done hair. Knowing an author, I felt like I had hit home. I bought the book, drove back to my hotel, and started reading. It gave me reassurance that I knew stuff considering that the use of a ball is perhaps the only consistency across golf and polo. I stayed in Florida for a week yet I never re-opened the book after that

> first day, first practice. But boy, that first outing as an independent mental coach was a scary one!
>
> (Ani Chroni, January 1997)

When starting as a sport and performance psychology (SPP) practitioner, one of the greatest challenges is applying what one has learned within the realities of the performance context. We all aspire to provide clients[1] with meaningful and feasible solutions as opposed to "canned approaches." Oftentimes, the translation of knowledge and tools provided in textbooks, journals, and courses to real-world interventions and applications can be challenging. While we all understand how important it is to comprehend theories that explain why/how things happen and theoretical best practices that aim to inform what can be done, balancing that with artful improvisation is the name of the game for what lays in the here-and-now of working within a client's setting. Case studies come in handy to bridge the gap between theory and practice, and can help one think creatively, out of the box, and on one's feet when it matters most. *Building Consulting Skills for Sport and Performance Psychology* is a collection of case studies (with solutions available separately in the Instructor Manual), aspiring to introduce developing practitioners, like students and early career professionals, to pertinent issues within a variety of performance environments while learning from the wisdom and experiences of professionals with perspectives and approaches from different consulting and cultural lenses.

Essential to student and practitioner development is an opportunity to translate theoretical concepts into action – the essence of the "scientist-practitioner model" – all while being mindful of the context, culture, and rapport required to implement effective interventions. *Building Consulting Skills for Sport and Performance Psychology* offers a unique opportunity for developing practitioners to integrate theory and practice in a collaborative, fail-forward environment. With supervisors, faculty, classmates, and peers, students and early career professionals will be able to work with challenging situations in order to integrate theory and practice, and slowly develop and refine their own consulting approaches and philosophies. Enhancing practitioner comfort with case conceptualization, intervention design, and personal reflection is essential to ensuring quality service provision during the graduate supervised experience as well as during post-graduation continued learning and growth.

In recent years, the use of case studies has gained popularity in the education and development of SPP practitioners (e.g., the development of *Case Studies in Sport and Exercise Psychology* by AASP and Human Kinetics). The case study approach (particularly when paired with case solutions) illustrates potential situations a consultant may encounter, offers insight into strategic decision-making, and reveals both effective and ineffective solutions to common problems within a variety of sport and performance arenas. Case studies as a learning approach work through stories, giving glimpses of others' real-life lived experiences, values, meanings, relationships, commitments, and ways of being and doing. We learn from stories as they speak to us louder than any other learning modality. According to Smith and Sparkes (2009), "the more stories we have access to, the more potential opportunities and flexibility we may have to live in different and more meaningful ways" (p. 280). The stories told in cases instruct, reveal, and inform us in unique ways; they help us make sense of context and allow us to share our understanding with others. To make a difference in the lives of clients, Mental Performance Consultants[2] need to be knowledgeable, skilled, creative, ethical, communicative, culturally sensitive, caring, empathetic, and, of course, solution-oriented. Yet, as in real-life, there is no one-size-fits-all approach, neither with stories told in cases nor with solutions to those cases. It is very common for a solution that works in one setting for one consultant to not work in another setting with the same or a different consultant, even if the settings have strong elements in common.

The original *Case Studies in Sport Psychology* (Rotella et al., 1998) textbook is now 24 years old, and while many of its cases are still relevant to this day, the time is right for a new collection of cases that speak to today's diverse and growing SPP environment. *Building Consulting Skills for Sport and Performance Psychology* brings together cases from around the world, highlighting various performance settings, levels, genders, cultures, challenges faced, and consulting and counseling approaches. *Building Consulting Skills for Sport and Performance Psychology* cases bring together authors from all continents in an effort to emphasize the diversity of not only performers and performer settings, but also consulting approaches and philosophies. Did we manage to fit all possible cases and authors in this book? We are sure that we didn't! Yet, we are sure that we diligently tried to fit in more than ever before as we intentionally searched for voices and cases that aren't regularly published or highlighted. In sharing so many different cases, an aim of ours is to step away from the conformity and singularity of the SPP discipline (Chroni & Kavoura, 2020; Ryba & Wright, 2005) and endorse multiplicity as we come to accept diversity and non-conformity as integral for moving forward as professionals and a discipline within the global and unequal world of today.

With this book, we hope to challenge developing practitioners to conceptualize each case from their own perspective, recognize and acknowledge the inherent diversity and the impact of that diversity on all involved (including the consultant), and integrate research and theory into identifying meaningful and feasible intervention strategies for the setting. Please note: *there is no one right answer to any of these cases*! Instead, the purpose is the deliberate development of the individual practitioner's consulting skills, philosophy, willingness to entertain alternatives, and honest self-reflection regarding consulting strengths and possible areas for improvement.

Building Consulting Skills for Sport and Performance Psychology: An International Case Study Collection comprises 27 case studies written by prominent applied practitioners representing a wide variety of performance venues. These cases are organized in six sections considering cases from traditional sport, high-risk occupations, performing arts, future trends in consulting, ethical challenges, and consulting competencies. The cases offer abundant opportunities to dive into other peoples' stories and, through progressive reflection, translate others' experiences into one's own opportunity for learning and growth.

Thought Questions

At the end of each case, you will find questions to help guide you through various considerations and reflections. We refer to these as the CEDI questions, and they are organized into four main areas:

(1) Case Conceptualization,
(2) Ethical Considerations,
(3) Diversity Awareness/Reflection, and
(4) Intervention Planning and Evaluation.

The importance of systematically conceptualizing the case cannot be overstated. Understanding presenting concerns allows a practitioner the opportunity to think through the best potential interventions suited for the specific context. Perhaps most importantly, it also provides an opportunity to recognize information that is still missing so readers can "know what they don't know" about the case.

Ethical concerns can (and will) arise in Mental Performance Consulting, and consultants will benefit from developing the self-regulation skills crucial to ethical decision-making (Aoyagi et al., 2012). Therefore, questions about unique ethical considerations for each case are included to help the reader develop ethical antennae to assist them in learning more about ethically engaging with the work.

Although superficially the crux of a consultant's work looks to be intervention planning and evaluation, without first accounting for diversity and culture in each case, interventions will be limited in their effectiveness. Each client, team, group, and context bring distinct cultures and identities. Without taking time to recognize, acknowledge, and accept multiplicity and uniqueness in cultures and identities within the consulting setting, a practitioner might not only be ineffective, but could potentially do harm. When thinking about diversity, think about the multiple "me's" (identities) we all carry with us. Gender, race, religion, and ethnicity are critically important identities, along with a wide array of other identities with which the performer(s) identify. What identities (social, role, and personal) seem to be present in each case, how could that be impacting the performer(s), and how might you as the consultant learn more about the extent to which the performer(s) identify with each identity (and how can you create a space where they can safely communicate other identities they might hold)? We paid purposeful attention to ethical challenges and diverse populations (race, culture, gender, religion, sexual orientation, disability, etc.) to challenge readers' cultural competence as we invite them to identify their own uniqueness in the world and how it impacts their attitudes, beliefs, understanding of clients who are different than (and the same as) them, empathic connection to potential clients, and ways of working with clients. Cultural awareness and reflexivity, culturally competent communication, and culturally competent interventions are the components of cultural competence practitioners need for engaging in cultural praxis (read more in Hanrahan, 2010; Ryba 2009; Ryba & Schinke, 2009; Ryba et al., 2013; Schinke & Moore, 2011; Schinke et al., 2009). As Chroni and Kavoura (2020) argued, while cultural competence components are taught in parts, "focus is not placed on synthesizing them at a higher level, that of bringing these together and embedding through cultural lens in a holistic view of the setting" (p. 232). The cases provided here give readers the context along with the space, time, and auxiliary tools to improve their cultural competence, to find unity and fluidity among those components.

Lastly, when considering potential intervention strategies, readers are also encouraged to think through how they will evaluate the effectiveness of those interventions. Just as practitioners regularly ask clients to engage in self-reflection to learn about their personal strengths and areas for growth, consultants benefit from engaging in the same work, which includes holding space for appraising the work by finding room for clients' voices and experiences to be heard.

Recommendations for the Developing Practitioner

> I was fresh out of school and trying to find my way as a young consultant when I got the opportunity to work with a team preparing for an international competition. While in hindsight it was a fantastic opportunity, at the time I was nervous and scared of what the outcome might be. I remember thinking I hadn't been this anxious about an opportunity to perform since my days as an athlete. I reached out to a friend from my college days, and we discussed the opportunity and the experiences I could bring to bear to do the best job possible. While all I saw were obstacles, he reminded me of past success and processes we had developed together to prepare and perform our best when we were consulting with our team/clients while in school. Our conversation also reminded me of what got me this opportunity, which was my hard work, preparation, and repetition in preparation for performance. At the start of the first session, I was anxious. For a minute I forgot all I had prepared, and then I remembered my process, took a deep breath, recovered, and the session went off without a hitch.

> I felt confident, the conversation flowed, and we went overtime. The session was a success and the start of a great working relationship with the players and coaches. I was so excited that I had to decompress in the parking lot. I remember thinking, "I can do this!" Once I was able to slow things down and recover, I called my friend, sharing the thrill with him, thanking him for his support, and offering my gratitude. At a time when I only saw obstacles and trouble, he was able to remind me of the process; my process, my experiences, my preparations, and my readiness to do this work! To this day I take my process, and his kindness and support with me in all the things I do.
>
> (Doug Barba, 1999)

This book is intended for graduate students with foundational theoretical understanding, and early career professionals looking to expand their understanding of various contexts and who are continuing to articulate their consulting philosophy. That said, we have no doubt that seasoned practitioners will also benefit from these cases as a chance to reflect on how their own work has changed and evolved, while also thinking through the complexities of various settings, what we can learn from different domains, and what knowledge we can transfer across settings and borders.

Although students might initially be frustrated that the "answers" are not readily available to them,[3] this decision was intentional. Learning from more experienced practitioners is inherently valuable. However, without first thinking through your own personal approach, students and early career professionals might be tempted to emulate others rather than carving their own path. As Ken Ravizza would often tell his graduate students, "You can't be me – you haven't failed enough yet." Or, as Oscar Wilde said, "Be yourself; everyone else is already taken." We cannot emphasize enough that there is *no one solution to any of these cases.* One of the reasons multiple practitioners were vetted to write solutions is to highlight the various approaches practitioners can take to this work based on their own identities, experiences, training, background, beliefs, philosophies, etc. We encourage you strongly to see yourself in each case and to pause, think, and then think again about your thinking (reflect). What are your initial responses? Where might those responses come from? What might you struggle with in a case? Where might your strengths be helpful? Use the CEDI questions to further your reflective practice, exploring each component in more depth and width, adding more lenses to your toolbox through which you examine a case. With every client with whom you work, you will never have all of the information at the start as a Mental Performance Consultant; how might you do the best work you possibly can with the information you have, and what steps can you take moving forward to gain a better understanding and do the best possible good within each specific context?

Recommendations for the Faculty/Mentor Facilitator

> Throughout my undergraduate psychology degree, I remember nearly every professor saying something along the lines of "you have no business entering someone else's head with the limited knowledge you learned in this course." So when I decided to go to graduate school for sport psychology, I didn't even bother to apply to any program that promised an applied experience. "I'm not ready," I thought. Although I was safely able to avoid applied work during my master's program, I wasn't nearly as lucky in my doctoral program. In my first semester, I was assigned to a very small office space, up 5 flights of stairs in the gym, and began work with one of the school's star athletes. She discussed losing passion and motivation for her sport, and I felt confident I could help her discover a path forward. I introduced a relatively simple idea: Make a list of "everything you love about your sport" on one side of the page, and "everything that would improve in your life if you stopped playing your sport" on the other side. She immediately bought into the exercise, and we had a great conversation. She promised to continue her list when we finished the session, and we planned to talk about what she discovered in the next one. As she left, I remember feeling incredible – "That went great! I can actually do this! Putting off the applied work until I had more knowledge under my belt was the right thing to do!" I raced down those 5 flights of stairs directly to my advisor's office. I couldn't wait to tell him how well it went. When I finished relaying my experience, he looked at me and said, "What are you gonna do if she quits?" My jaw hit the floor. What if, in one day of work, I had led the star athlete to quit her team? My advisor just laughed at my reaction and shrugged his shoulders. He was waiting for an answer.
>
> (Sarah Castillo, 1996)

To those using this book as a teaching tool in a classroom or mentorship setting, thank you. We firmly believe in the power of learning through case studies and narratives and are confident this resource will be helpful for your students and advisees. In the Instructor Manual, you will find three solutions for each case presented in the Case Study Book: one solution written by the original case author outlining the steps they took in their work in that setting, and two other solutions from practitioners who did not work in the original setting but were willing to share their reflections on how they might have responded. Before sharing the solutions with students, we encourage you to allow students time – both individually and in small groups – to read through the case and respond to the CEDI questions. This might lead to more in-depth discussions with you and in larger groups. After discussing and debating their own approaches, it might be helpful for readers to engage in group learning and discuss their own approach with that of others, which can then lead to further discussions about what might have been missed, what they would change, and what they would not change. Our hope is that these cases help students transition from theoretical knowledge to conceptualizing how to apply that knowledge, along with recognizing the diverse complexities of real-world performance domains.

Embarking on a Journey of Stories

Every journey starts with one step; you can either start with the first case of this book or pick one in a performance category that interests or challenges you. You guessed right, *there is no one right way to go about it*, the first step is simply to start reading and reflecting on the stories that unfold in the next 196 pages. The cases presented in *Building Consulting Skills for Sport and Performance Psychology* capture a range of perspectives giving readers the opportunity to gain a greater understanding of the performance arena, sport, context, and/or challenges at hand. While starting and growing as a SPP practitioner takes knowledge, skills, and experience, it also takes a little bit of courage as we continuously expose our imperfect professional selves. We hope that learning with and from each other will ease many steps of the journey.

> Being in a master's program with an applied focus, I was placed with a team early in my graduate training. Although I was grateful for the trust shown by my mentors and advisors, I did not feel ready (though, upon reflection, I don't know that I ever would have felt "ready"). The coach could not have been more bought into mental training – she had worked with consultants in the past and knew the value of the work. This buy-in was so strong, she scheduled a 2-hour session for me with the team after a Saturday practice. I couldn't believe it! I had been learning about the difficulties of getting buy-in from coaches, and here I had a coach who was dedicating 2 full hours to mental training! I didn't even question if that was a good thing or not. I went into that session nervous but excited. I had a plan, and I knew it was going to be great. I quickly realized that I had not accounted for the context around the session. The team was wrapping up "hell week" – they had been in multiple practices each day all week, and that Saturday morning they had started practice at 7:00am (early for anyone, much less college students); our session was at noon – after 5 hours of practice and during lunch time. By the time they got to the session, they were completely exhausted. Unfortunately, being so new to the work (and my Type A personality taking over when I am nervous), I decided to stick with my plan and keep them for nearly the entire time. That meant moments intended to be full of rich discussion were instead filled with silence or short responses; the workshop turned into more of a lecture than a participatory discussion. I am also fairly

certain that very little (if any) of the information I shared that day actually stuck. Luckily, I had the opportunity to debrief the experience with my classmates and mentors soon after. We talked in-depth about the pros and cons of that length of session, the limitations of coming in with a hard-and-fast plan, and changes I might make moving forward. Although that might be one of the worst sessions I have led in my career, the opportunity to learn from it (with the help of many trusted others) served invaluable.

(Chelsea Butters Wooding, January 2011)

Notes

1 The authors acknowledge that not all people with whom we work become long-term, paying clients. Our use of the term "client" throughout represents all with whom we interact, work, and communicate, regardless of the length of time the relationship exists.

2 We deliberately use the term Mental Performance Consultant here based on the certification administered by the Association for Applied Sport Psychology, which is the oldest one worldwide. We understand this title might not accurately represent the training and work of practitioners in other countries as other organizations, like FEPSAC and ISSP, have recently also initiated certification programs.

3 Solutions are only available in the Instructor Manual; solutions are not included in the Case Study book. If students only have access to the Case Study book, they will not be able to readily access the solutions.

References

Aoyagi, M. W., Portenga, S. T., Poczwardowski, A., Cohen, A. B., & Statler, T. (2012). Reflections and directions: The profession of sport psychology past, present, and future. *Professional Psychology: Research and Practice, 43*(1), 32–38.

Chroni, S., & Kavoura, A. (2020). Cultural praxis. In D. Hackfort & R. J. Schinke (Eds.), *The Routledge international encyclopedia of sport and exercise psychology: Applied and practical measures* (Vol. 2, pp. 227–238). Routledge.

Hanrahan, S. J. (2010). Culturally competent practitioners. In M. B. Andersen (Ed.), *Routledge handbook of applied sport psychology* (pp. 476–484). Routledge.

Rotella, B., Rotella, R. J., & Boyce, B. A. (Eds.) (1998). *Case studies in sport psychology*. Jones & Bartlett Learning.

Ryba, T. V. (2009). Understanding your role in cultural sport psychology. In R. J. Schinke & S. J. Hanrahan (Eds.), *Cultural sport psychology* (pp. 35–44). Human Kinetics.

Ryba, T. V., & Schinke, R. J. (2009). Methodology as a ritualized Eurocentrism: Introduction to the special issue. *International Journal of Sport and Exercise Psychology, 7*, 263–274.

Ryba, T. V., Stambulova, N. B., Si, G., & Schinke, R. J. (2013). ISSP position stand: Culturally competent research and practice in sport and exercise psychology. *International Journal of Sport and Exercise Psychology, 11*, 123–142.

Ryba, T. V., & Wright, H. K. (2005). From mental game to cultural praxis: A cultural studies model's implications for the future of sport psychology. *Quest, 57*, 192–212.

Schinke, R. J., Hanrahan, S. J., & Catina, P. (2009). Introduction to cultural sport psychology. In R. J. Schinke & S. J. Hanrahan (Eds.), *Cultural sport psychology* (pp. 3–11). Human Kinetics.

Schinke, R. J., & Moore, Z. (2011). Culturally informed sport psychology: Introduction to the special issue. *Journal of Clinical Sport Psychology, 5,* 283–294.

Smith, B., & Sparkes, A. C. (2009). Narrative analysis and sport and exercise psychology: Understanding lives in diverse ways. *Psychology of Sport and Exercise, 10*(2), 279–288.

Section 1

Cases from Traditional Sport

1 Youth Sport

"Just Don't Lose"

Karen Lo

Hong Kong

"Hi ... so, this is my daughter, Cassie. She's a very promising squash player. She placed 13th at the US Squash Open a month ago. We were expecting to be in the top eight," Cassie's father, David, explained. "She's quite bummed. She loves squash a lot and wants to do really well. Right kiddo?" Cassie nodded, but she did not meet her father's eyes. "I kind of want to explore how you might be able to improve her game. That's why we're here anyway," he finished. Cassie glanced at you shyly but didn't say a word. You immediately felt the tension in the room.

David left the room after some time, and Cassie started speaking in leveled tones. "I'm not confident when I'm on the court, especially when I'm playing against someone I know I shouldn't lose to," she admitted, diving right into the problem. "My dad also thinks it's a confidence issue. They know I get scared when I'm playing against someone younger and weaker. He says I can't afford to lose those games."

During the session, you started to learn a lot about Cassie's story. Cassie goes to a local school and recently turned 14. She started learning squash at the age of six, and instantly fell in love with the speed of the game. Over the past years, she made a small group of friends within her squash community, trained hard, and is easily one of the best in her age group. She used to be on the track and tennis teams when she was in primary school, but gave both up so that she could focus fully on squash. She also spends two hours each week learning the flute and the piano.

Cassie comes from a high-achieving family. Her parents were both educated in the United States, and have high hopes of Cassie entering an Ivy League University with her decent grades and squash performance. Cassie's older sister plays squash and softball and has recently been accepted to Tufts University in the fall. Knowing that there are very few people who play squash in Hong Kong (HK), her parents decided that squash could become Cassie's edge to entering a good university, which could lead to a better future and more career opportunities. "I've worked in the US for eight years. Companies love hiring student-athletes because they're so disciplined and tough, and I know you are capable," her father mentioned to her once. Cassie respected her parents, knowing that they were both hard workers. She had always been obedient, and knew early

DOI: 10.4324/9781003089629-2

on that the path her parents decided for her would not only maximize her chances of representing HK to play at an international level, but also of being accepted to a highly-ranked university. All her life she had learned the importance of getting into a good university for a better future, so she was quite accepting of the idea of having her parents plan her future for her.

When Cassie turned 11 and began secondary school, she realized that her schedule became increasingly hectic. Apart from taking regular weekly piano and flute lessons, she was playing squash for 11 hours a week. Her parents noticed her vast improvement in squash, and made her quit track and tennis. Her entire week would be spent on school, squash, flute, piano, and tutoring classes. She trains heavily under the supervision of her two squash coaches. One of them is an older "skills" coach who has been with her since she first picked up a racket. The other is a younger, energetic coach, and a former HK Team player, who also acts as her fitness trainer. Cassie does private and group sessions with both of them. David picks Cassie up from practice and chats with the coaches to check her progress. Her younger coach noticed Cassie's speed and agility and believed she could be more successful, and suggested to David that they increased her training load.

David agreed to it immediately, as he was pleased to know that Cassie had the potential to do even better. After discussing this with Cassie's mother, they both agreed to add another private coach and an additional group class to her already busy schedule. Cassie agreed to the plan. She started training for 13 hours a week.

As time went by, her squash performance rocketed, and she managed to rank third in the U12 category. She quite enjoyed the feeling of being at the top; it was euphoric. At that time, she was good friends with another stocky girl named Jenna who was fourth in her age group. Jenna was nine months younger. Cassie had always beaten Jenna, as Jenna could not run as fast as her, but over the past five months, she slowly realized that she would secretly feel relieved if Jenna wasn't playing the same match as her. She also reported feeling nervous whenever she was seeded against Jenna or anyone who ranked below her. Feeling like she needed to keep up with her ranks, Cassie felt pressured to win every match she entered. She admitted feeling that her confidence would waiver, her mind would go blank, and she wouldn't be able to think.

Cassie recalled one of the matches she played half a year before the US Open. It was against a girl who was only 12 years old. Before Cassie stepped into the court, her coach said to her, "She doesn't have as much experience as you Cassie. You should be able to win this." Cassie had complete trust in her coaches, but wondered for a moment what her coach would say to her if she didn't win in the end. She could feel her heart thumping loudly. Her dad echoed the coach's comments. She walked in feeling stressed, and there were a few points she missed. She won the match by a narrow window. "I think maybe my Dad just thinks it's fine to repeat what the coach said," she said. "I don't feel confident when he says it though. He doesn't even play squash." This was the start of the nightmare. She noticed that she would feel relieved when matches were canceled, or if Jenna or the other lower-ranked girls did not take part.

When she broke for summer vacation, her parents decided to place her in a three-week training camp in southern Malaysia so that she could receive higher quality training. Her parents had a few connections with the coaches there. "It was a relief to be able to train with other people once in a while. Sometimes being with people I know scares me the most," she added uncomfortably. She learned new tactics and strategies, and thoroughly enjoyed her time playing with people she barely knew. She was happy nobody knew who she was, had no expectations of her skill level, and did not judge her if she didn't train well. However, she knew with a sinking heart that she would have to face the same opponents again back at home.

Shortly after her trip, she started school again, and her schedule was just as busy as ever. Her coaches signed her up for the HK Close before she flew away for the US Open. She saw Jenna there and was surprised to see that Jenna had lost a bit of weight and was much faster on court. "I left HK, so I had no idea where she went for summer break," she said. She sat on the bleachers to watch Jenna play against another opponent, trying to get a better sense of her current style. Jenna eventually won and Cassie felt a tinge of apprehension. She was beginning to feel her heart pounding. Jenna won. *Jenna won.* This meant she needed to play against her now. She started to panic. *I can't lose*, she said to herself. "You've beaten her multiple times, Cassie," said her dad, trying to sound reassuring. It didn't help. She didn't feel ready at all.

Don't lose, she reminded herself. She trailed behind Jenna and walked into the court. She gave a quick glance at Jenna, who looked composed and confident. She felt her hands shaking. *Get a grip*, she muttered between her teeth. Her rallies were short. Her returning serve felt bad. Everything felt wrong. She tried to fight back and gain control over the ball, but couldn't. Her knees shook, and she didn't know or remember what happened. She just knew she lost.

She didn't want to head back to where her dad sat. She needed time to think. She avoided his gaze and ran straight to her friends, hoping to find a moment of distraction. Her friends tried to cheer her up by saying that she was just unlucky. She felt exhausted. She slowly made her way back to where her dad sat, hoping her dad wouldn't say anything.

"What happened just there?" her dad frowned and asked. She was trying to hold back her tears, and simply uttered, "I don't know. I'm tired." Her dad stopped saying anything in the car, but picked up the conversation again over dinner. "We've spent so much money on your training and overseas tournaments. Where has the money all gone?" It wasn't the first time her dad said that. He then started critiquing her game. He sounded exasperated.

She stayed silent, but secretly felt annoyed and a little agitated. The US Open was just around the corner; she was quite worried as Jenna was also going to be there. What if she was seeded to play against her again? She would lose to her twice in a row, and that would be a terrible blow to her own ranking. *I can't lose to people who are younger than me*, she thought to herself angrily. If she kept losing, it would ultimately jeopardize her chances of getting into a good college. It wouldn't look good on her profile either.

She flew to New York with her mother and experienced a rather hectic two weeks. Her seeding was higher than Jenna, so she had to face a number four seeded player in the top 16 round while Jenna played against a number eight Egyptian player, Aliya. Aliya didn't play well, which led to Jenna advancing to the top eight. Meanwhile, Cassie lost and had to go to the consolation round to fight for 9–16th. During this round, Cassie had the chance to play against Aliya. Unfortunately, Aliya did exceptionally well in that game, and Cassie lost quite badly. She was placed 13th in the tournament.

> "It was quite humiliating, losing to that girl," she muttered.
> "You felt humiliated losing to Aliya?" you tried to confirm.
> "Yes, because Jenna won when playing against Aliya. So, I've actually lost to Jenna too."

The names of the characters in the story have been changed to respect the real athlete's confidentiality.

Case Conceptualization

- What information do you need prior to beginning to work with Cassie in order to enhance your effectiveness?
- How much does Cassie's upbringing (family background) play into her current state?
- How does she feel about having to focus on a sport purely to get into a good college?
- How does her current training program (primarily scheduled by her parents) impact her psychologically?

Ethical Considerations

- Do the facts in this case suggest any "red flags?" That is, do you notice the need to consider clinical referral, limits of your consulting competency, competing stakeholders, etc.?
- She is under 18, and her parents are paying for your services. The father would like to be occasionally informed on how his daughter is doing. What would be things that can be shared with the father, since a fair amount of the conversation between you and Cassie revolves around her dad?

Diversity Awareness/Reflection

- What are your initial thoughts regarding the culture(s) involved in this case? These may be related to the performance arena or the individual performer.
- What might you want to learn about the parenting culture/environment in HK that could help you better understand some of her underlying issues on the court?
- How important are the parents' values on her values within this sociocultural context?

Intervention Planning and Evaluation

- Identify the potential issues that may be affecting Cassie (these may be personal and/or performance-related). From your list, prioritize the top three issues that you believe are having the most immediate impact.
- What ideas or interventions have you identified that might help Cassie, and how would you prioritize them?
- How does she determine what success in squash is?
- What are her thoughts about her own life so far? Is she satisfied with it?

2 Collegiate Sport

"The Stoic Collegiate Swimmer"

Poppy DesClouds and Natalie Durand-Bush

CANADA

Introduction

This case involves a swimmer named Marina, who is currently a first-year collegiate athlete with whom you, a Certified Mental Performance Consultant (CMPC) with a sport science background, have been working on and off since she was 15 years of age. Marina competed for her local swim club for four years and then transitioned to her collegiate team after receiving a scholarship to attend a top NCAA Division 1 college. Part of your initial consulting work with Marina is included in the introduction as it is important background information leading to the issues that Marina is experiencing in her first year as a college athlete, for which you need a solution.

Marina has always been a mature, powerful swimmer with tremendous potential. In Marina's view, she started her sport late and had to work hard to "catch up" to those around her. Marina began swimming when she was four years old but only started competitive swimming at fourteen. Outside of swimming, Marina is a dedicated student who enjoys spending time with her family, friends, and boyfriend. From an academic standpoint, she has contributed to school committees and has always worked hard to maintain a high great point average. Although she will not speak to her friends or parents about it, Marina dreams of having an outstanding collegiate swim career and graduating with honors from her college.

Club Level

At the club level, Marina trained six days per week under the direction of a young female coach, with whom she formed a unique bond and friendship. Her teammates knew Marina as having untouchable force, endurance, and performance presence. Marina was not fully aware of her reputation at the club level, and although she liked being a role model, she felt pressure to perform and maintain a certain status for her team and coach.

Marina has had a deep competitive drive since she began competing. At small meets, she often swept out her competitors, performing strong, consistent, and seemingly effortless races. At larger competitions,

DOI: 10.4324/9781003089629-3

she raised the bar and was motivated by her competitors and enjoyed watching and competing against the swimmers she regarded as better than herself. Over the years, Marina has been methodical, setting and meeting pace targets and keeping her focus on her personal goals. She has not always attained winning times, but has consistently performed personal bests and has kept herself within the top of the pack. She has been particularly successful in her distance-free swims, which she has enjoyed more than any other race.

In terms of your consulting work with Marina at the club level prior to her becoming a collegiate athlete, you had regular sessions with her as part of the mental performance program you were implementing with athletes at the club. This allowed you to develop a trusting and open relationship with her. However, during the year prior to her transition to the college setting, you observed that Marina often disregarded her self-care and recovery. She had a persistent back injury, and she often came to practice overtired and depleted. Marina also became frustrated about being the best swimmer at her club. She felt she had no one to look up to for motivation.

You heard whisperings from other swimmers that Marina threw up before her 1500 m free swim during her last competition. You witnessed Marina throwing up once when she was training, but she chalked it up to fatigue, swallowing pool water, and not eating enough. The next time you saw Marina at a competition, you gently asked her if she was feeling OK. She disclosed that she was throwing up mid-way through each meet, for the last four competitions. Marina was confused by the behavior, as she was not making herself do it, and felt like it "snuck up on her." Marina reported that it usually happened after she had completed a couple of races and had "sussed out the competition." When her energy peaked, she got a rising feeling in her belly and throat, and she threw up quickly and discretely. Marina felt that throwing up was a sign of nervousness, as she felt better right after vomiting. In your discussion with her, you acknowledged that throwing up is not an uncommon component of performance anxiety, but it can be an uncomfortable and non-optimal coping strategy.

Over a couple of meetings, you worked with Marina on (a) acknowledging, accepting, and reframing the pressures that made her feel nervous (e.g., pleasing her audience, representing her club, the potential of throwing up), (b) acknowledging and working with her stress response, (c) activating her parasympathetic system, to achieve a sense of calm and control, and (d) managing her energy expenditure before and between races (Perry, Shaw, & Zaichkowsky, 2011). Together, you and Marina created a scale based on her individual zone of optimal functioning and identified her optimal pre-performance activation (Hanin, 1997). You collectively brainstormed ways that Marina could achieve and manage these levels, including check-ins, breathing, relaxation, self-talk, and various physical and mental cues. You continued to observe Marina's physical and mental state and checked in with her regularly.

Marina immediately began to practice these strategies at training and competition, and her feelings of needing to throw up completely subsided. She reported being able to manage her pre-performance state much better, particularly accepting, normalizing, and working through her body's response to pressure. Although she felt tired after her races, her energy was much more manageable and consistent. You observed Marina in good spirits, showing more energy and focus at training, with less frustration. She finished the competitive season strong, and she was content with her performance.

Collegiate Level

Now that Marina is competing on the collegiate swim team, the environment is entirely different, but she loves it. She is motivated by the high-caliber swimmers, the challenge of a new coach, and a more intense training context. Her new coach is strict but kind. Marina gets along well with her, and her coach appreciates Marina's strong work ethic and ability. Marina is performing so well at training that she "forgets about" her mental training practice and routines from the years before.

The training schedule and intensity of practices cause Marina's back injury to flare up. She feels overwhelmed by her pain but continues to train hard and hides her discomfort. Some days after training, Marina can hardly walk, but she continues to make it to every practice and every class. She gets her schoolwork done to a high caliber and makes new friends inside and outside of swimming. She also stays in touch with her high school friends and old coach.

However, after a few months into the season, Marina sits down with you and immediately begins to tear up. She tells you she's been feeling depleted and stressed, struggling to balance school, swimming, and her social life, which is negatively affecting her performance. Also, she reveals that her body "sucks." She is working hard with her physiotherapist but is still in a lot of pain. You give Marina space to cry, and she quietly discloses that some evenings she has started vomiting again, and it's happening at home now. You validate Marina's feelings and ask whether she has been feeling ill or throwing up for any other reason. Marina says she doesn't know. You inform her that she is in a non-judgmental space and that you are there to listen if she does feel the need to disclose anything. Marina nods, wipes her face, and starts to discuss more details about being stressed, exhausted, and frustrated, especially in the presence of her friends. She takes a pause and then lets you know she's been throwing up consistently after seeing her friends. Eventually, she opens up completely and explains that she has been making herself throw up.

Marina explains that on days when she is extremely tired and stressed and sees her friends, she starts to feel under-confident in her abilities and body image and makes herself throw up. She explains that throwing up gives her a sense of relief. Marina tells you she knows this isn't an ideal or helpful behavior but believes that she is using it as a necessary coping strategy. Marina admits that she has wanted to tell you about this for a while, but she was unsure and hesitant to open up. She isn't throwing up at swimming, and she is extremely worried that people will think she is weak. Marina stops during the conversation from time to time and says things like, "Don't worry, I'll just make myself so busy that I can't do it anymore"; "I'll just force myself to stop throwing up." Marina is unconvinced that she needs "help," especially not from someone who will "just tell her to love her body." She says she is motivated by being disciplined and hard on herself. Marina has an intensive competitive season and a tough academic year ahead, and she is feeling both excited and scared. She wants to focus on these priorities and restoring her performance, much more than she wants to address the immediate issue of using vomiting as a coping strategy.

Reflective Questions

Case Conceptualization

- What information do you require prior to continuing your work with this swimmer in order to optimize your interventions and effectiveness as a CMPC?

Ethical Considerations

- Are there any ethical issues you should consider (e.g., competence, consent)? More specifically, should you consider a clinical referral given the limits of your consulting competency? Should you continue working with the swimmer if you refer her to a clinician?

Diversity Awareness/Reflection

- What are your initial thoughts regarding the culture of college athletics and the culture of swimming that could impact this case? What are the potential contextual barriers and facilitators?

Intervention Planning and Evaluation

- What are the top three important issues affecting the swimmer that you should prioritize (e.g., personal, performance, mental health issues)?
- What compounding stressors could be influencing this swimmer?
- What unique aspects of this swimmer should be addressed in your consulting work with her?

- Are there clinical mental health concerns, and if so, what are they?
- Which aspects of this case are suited for interventions by a CMPC and or for a mental health practitioner (e.g., clinical counselor or psychologist)?
- As a CMPC, how can you make a difference with this swimmer? What advantages do you have?

References

Hanin, Y. L. (1997). Emotions and athletic performance: Individual Zones of Optimal Functioning model. *European Yearbook of Sport Psychology*, 1, 29–72.

Perry, F. D., Shaw, L., & Zaichkowsky, L. (2011). *Biofeedback*, 39(3), 95–100.

3 Club Level Sport

"The Struggling Dual Career Support Provider"

Koen De Brandt, Paul Wylleman, and Simon DeFruyt

BELGIUM

Stella, at the age of 28 years old, is a novice dual career support provider at a European university. For the past two years she has been assisting student-athletes in their dual "university and high-performance sport" career pathway. Stella is no stranger to the university where she currently works. She was a well-known student-athlete herself, combining a professional beach volleyball career with studies in sports science. After graduating at the age of 24, Stella continued to play beach volleyball at the international level. Sadly, at the age of 26, she suffered a severe hip injury during a continental championship and was forced to end her professional playing career. Shortly after the incident, Stella underwent surgery.

While the abrupt end of her career came as a big shock to her and those in her closest circle, Stella also came to view it as an event that also opened new avenues for her. Within three months after surgery, she successfully applied for two part-time jobs: one at her alma mater in the position of coordinator of the brand-new student-athlete department, and the other as a coach at a regional beach volleyball academy (training center). Currently, Stella's time investment between the university and the academy is about 50 50. At the academy, she coaches a group of twelve 15- to 18-year-old talented beach volleyball players over four practice sessions a week and additionally acts as the physical trainer for players' strength and conditioning sessions – two per week. At the university, Stella's job description as a dual career support provider consists of improving the support services provided to an average of 95 Belgian student-athletes every academic year. All student-athletes are officially recognized as talented or elite athletes by their respective sports federation. During Stella's time as a student at the university, services for student-athletes were nonexistent. She had to take the initiative in requesting and receiving support from her professors in order to successfully complete her studies while at the same time training and competing in beach volleyball at the highest level internationally. Other athletes who also studied during their athletic careers also had a hard time, and often complained about the support services provided or lack thereof. Eventually, Stella became one of the few athletes who successfully completed her degree.

DOI: 10.4324/9781003089629-4

Upon starting with her university job, Stella realized that support services for student-athletes at the university had not improved since she graduated, and that she had to start developing them from scratch. During her first two years she developed and improved her connections with professors, faculties, and study counsellors, and managed to establish a strong and well-delineated central organization providing flexibility in the way student-athletes studied. The organization acts as a central point of contact for student-athletes and aims to provide maximum academic flexibility in order to enable athletes to train and compete in their own elite sport environment (i.e., at their club or federation) outside the university. She was very proud to have introduced this flexibility for student-athletes in the exam regulations at the university. Once this system was in place, student-athletes had permission to be absent from their lectures, could change from one study group to another, as well as change exam dates when these coincided with competition days.

Stella's work with the student-athletes themselves has mainly been academic and organizational in nature. Before the official opening of each academic year, she meets with all student-athletes in person for a one-hour intake session with the aim to discuss and establish an individually tailor-made academic route, to identify potential overlap between academic and sport activities, and to gain an overview of the flexibilities each student-athlete requires with regard to attendance and exams. During those two academic years, the groups of student-athletes have been very diverse as they represented 29 different sports (e.g., athletics, basketball, field hockey, gymnastics, volleyball, swimming), enrolled in 25 different study programs (e.g., economics, engineering, psychology, physiotherapy, sports science) and 8 faculties.

The time of the year that Stella holds the student-athlete intake sessions at the university coincides with new season preparations at the beach volleyball academy, hence the start of the new academic year is a very busy period for her. But as Stella is a busy bee, she does not mind working hard, and feels that she receives a lot of energy from her conversations with the student-athletes. She has noticed that the time she invests in the student-athletes before the start of the academic year pays off during the year, as student-athletes start the academic year better prepared which in turn results in fewer unexpected requests during the year (e.g., last-minute requests to reschedule exams).

Once the academic year starts off, her workload decreases and the work becomes more administrative in nature. This mainly includes sending out and following up requests to professors and maintaining regular contact with the student-athletes to ensure that appropriate support is provided to them in view of their academic needs. She also invests a lot of time and energy in establishing and maintaining a trust-based relationship with all student-athletes who really appreciate her efforts and her continuous availability.

Since Stella started working at the university, she has also been tracking the academic performances of student-athletes. To her delight she has found that their grades are steadily improving and that only a few student-athletes eventually drop out of university. Stella is very proud of the progress she has made since she started providing support services to student-athletes. Indeed, the results of her efforts have not gone unnoticed. James, Stella's supervisor and head of the Athletic Department, several of her colleagues, professors, and coaches have congratulated her on the work done and the achievements made so far. Stella is extremely pleased with this positive feedback which motivates her to continue optimizing the support services for student-athletes as well as to keep on building the university's reputation as a student-athlete-friendly university. After ending the second academic year very satisfied, Stella decided to take a two-week break during summer to rest and kick off her third year at her university job with renewed energy. Finally her hip had healed completely, and she could play in some local beach volleyball tournaments and get back in shape. Stella strongly felt that life was good!

Right before she left for a well-deserved holiday, Stella sent an email to all student-athletes wishing them a good summer and asking them to complete an anonymous online survey about the services she provided to them. And off she went to rest and play some beach volley!

Stella Starts to Struggle ...

Upon returning from her holiday, Stella started to analyze the feedback she received from the student-athletes. While much of the feedback was positive, still 55% of the student-athletes indicated there was still room for improvement. Student-athletes reported that they often struggle with challenges specific to their dual career pathway, and that they would benefit from "some mental stuff" to cope with these challenges. For example, several student-athletes reported lack of motivation to continue with the combination of sport and studies, failure with planning and prioritizing, difficulties in coping with the concurrent pressures from the sporting and academic domains, limited social contact with peers because of their overloaded schedules, and lacking long-term vision. On top of that, Stella received a few phone calls from parents informing her on issues their children were facing, suggesting that some more "specialized" support from the university would be appropriate. While Stella recognizes some of the challenges from her own time as a student-athlete, she is somewhat surprised at the number, the diversity, and the intensity of the challenges the student-athletes report that they struggle with. Doubtful about how to proceed with this feedback, Stella decides to schedule a meeting with her supervisor.

Meeting with Supervisor

Stella's meeting with her supervisor (James) was rather rushed and not very enlightening. While James stressed that he was very happy with the organization Stella had set up during the past two years, he also made it very clear that the needs expressed by student-athletes and their parents required urgent follow-up. James instructed Stella to take several initiatives as soon as possible to support the student-athletes with their motivation, planning, and stress management. He suggested that she take a look at some "competency assessment and development" initiatives, which was something he had heard of and which appeared to "fit with the scope of a modern university." For James, being a modern university would ensure that the university could maintain its positive attitude toward athletes. Stella asked a few questions about exactly what these initiatives should look like; however, his response was that he had actually no experience with this kind of support and that she should ask around. James also emphasized that, if need be, there was actually no budget to hire another half-time support provider, and that she needed to squeeze all of it into her current workload. While feeling pushed to take action, yet at the same time feeling left in the dark, Stella was still positive and confident that she would do her best for the student-athletes and said to James as she was leaving his office that she would develop these ideas further.

Two Weeks Later

Two weeks later, Stella was completely stressed and overwhelmed. She searched and searched the internet and found tons of information on student-athletes' psychological and psychosocial development. While the topic is of great interest to her, she was not very familiar with it. During her career, Stella never consulted the sport psychologist or student counsellor, let alone having experience in providing this kind of support herself. She had no clue where to start and time was not on her side.

As the academic year is about to start, Stella knows too well that at this time of the year, the intake sessions with student-athletes consume most of her time and energy while working for the university, and at the same time her beach volleyball academy work demands a lot of attention as the start of the season approaches. On top of her part-time jobs, the national beach volleyball team invited her to train with them again twice a week. Stella is very motivated to improve her services to student-athletes, but she is also very concerned that increasing the time she spends mentoring student-athletes will jeopardize her other activities. In great need to know how to proceed with improving her support services to the student-athletes, Stella decides to call you, the head of the student-athlete department at another university, whose name she came across during her search on the internet. How can you help Stella develop initiatives which could, in light of the feedback she received, optimize the services provided to student-athletes at her university?

Thought Questions

Case Conceptualization

- What information do you need prior to beginning to work with Stella in order to enhance your effectiveness?
- If motivation is not holding Stella back, what specific factors might be playing a part in her current struggles?
- Stella is part of a larger organization. What is important to consider when working with her case?

Ethical Considerations

- Do the facts in this case suggest any "red flags?" That is, do you notice the need to consider clinical referral, limits of your consulting competency, competing stakeholders, etc.?

Diversity Awareness/Reflection

- What are your initial thoughts regarding the culture involved in this case? These may be related to the performance context or the individual.
- What contextual knowledge of the European dual career culture do you have to help you in this setting? What contextual pieces are missing that you would like to know more about?

Intervention Planning and Evaluation

- Identify the potential issues that may be affecting Stella (these may be personal and/or context-related). From your list, prioritize the top three issues that you believe are having the most immediate impact.
- Stella is working with a fairly large and diverse group of student-athletes. How will you ensure that all student-athletes can be involved?
- Stella is very eager to develop herself. How can you ensure that your intervention also addresses her personal needs?

4 National Level Sport

"A National Team in Crisis"

Carolina Lundqvist and Henrik Gustafsson

SWEDEN

The Team, as a Group

Today, the national team consists of a relatively stable squad of elite athletes who are all competing at a high international level in martial arts. This particular branch of martial arts has relatively few practitioners in the country but when some of the athletes perform well in major competitions and championships they are recognized in the national media. Looking back at the national team's history, international successes varied over periods. At times, there have been only single national performers with the capacity to seriously compete for medals at the international level. At other times, a greater number of talented performers successfully represented the country internationally. The athletes that make up the team today grew up training and competing early on for different, and sometimes, rival clubs which are spread around the country.

Five years ago, the core national squad consisted of only four athletes of roughly the same age, who had trained together for a long time with the same coach. They knew each other well, both inside and outside sport. They pushed each other and collaborated during training sessions to develop athletically, and because of different weight classes they rarely had to compete in national team selection trials or against each other at international competitions. At that time, their performances were outstanding at the national level – a performance gap existed between them and other athletes – and they were seen as young and promising athletes. Today, while more athletes make up the national team, these four are the most experienced ones with the greatest medal record internationally, and receiving the most media attention over the years. For a long time the people around them, including national media, considered the four as martial arts "stars" and their spots with the national team were relatively safe.

The Team, in Crisis

In the past two years, however, things changed. A new generation of talented juniors reached the senior level, some of them challenging the older athletes at team selections for a spot to represent the country at international

DOI: 10.4324/9781003089629-5

competitions and championships. At the national level, the performance gap has narrowed rapidly between the old and new generations – the older more experienced and the younger upcoming athletes. The success of the younger group of athletes has caught the attention of media and sponsors, giving them greater coverage by the media and more sponsorship events. The increasing attention to the new members of the national team has not been well received by the older athletes, who have been fighting for sponsor support and media attention over the years.

The national federation welcomed the re-birth of the sport; the growth of new, young athletes with a promising future ahead whom in a rather short period of time have built the potential to compete for medals at international championships. Seeing this positive development for the sport, the federation encouraged and supported the young athletes to move to the same town together with the older athletes to receive the best possible training opportunities. By doing so the young performers would also have the opportunity to be coached by recognized and highly qualified coaches, experienced in raising international champions. So the "core squad" of the national team increased from 4 to about 10 athletes, ranging in age from 20 to 30 years old.

Athletes on the national team are expected to practice most training sessions together under the guidance of one head coach and the assistant coaches. As such, they spend a lot of time together in training, while also living close to each other in apartments provided by the federation. Having only recently moved to join the national team's training site, the younger athletes have a limited social network outside sports in their new hometown. Over the past two years, as the national team gradually expanded with younger athletes joining, a general feeling of dissatisfaction started growing within the team. Conflicts and disagreements have been noted among team members and the team's climate is rather tense with an underlying sense of irritation in the air. Two camps have formed within the team; one consisting of the four older, more established athletes and one consisting of the younger, developing athletes who recently joined the team.

The camp of older athletes has gotten into the habit of gathering in the locker room after training sessions to vent their negative feelings. They complain in tandem about the coach not providing them with sufficient support and the training sessions not being good enough for them since the younger athletes joined the team. They complain about the younger athletes' behaviors, for example, not paying attention during practices and not acting professional enough as elite athletes, and that the younger athletes are favored by the coach. Each time the group of older athletes gathers, which is basically after each training session, the discussions get more and more heated; their anger escalates and the perception of being unfairly treated intensifies. None of them, however, wants to talk with the coach directly about these views and critiques; they think the coach should understand these problems anyway. For them, "It is too obvious." They are now adamant that the younger athletes align with the coach to get attention and favoritism for professional gain. In their view, the young "newcomers" need to "back off" and adapt to the "unwritten rules" that have always existed and applied.

Alexis is one of the older athletes who has been on the national team for 10 years. As a youngster herself, Alexis enjoyed rapid athletic development and at an early age performed successfully in several international competitions. Back then, she was viewed as "young and promising" by people inside the sport milieu and the media. In recent times, however, Alexis has experienced a performance plateau, while performances at the international competition level have improved. While Alexis has still managed to perform satisfactorily at international competitions and championships, the great successes seem to be a thing of the past and her self-confidence has decreased over time. She has become more and more stressed about her situation and afraid she might lose the "star" status she has held over the years. While worried, she tries hard not to admit it to herself and others, and pretends that everything is under control. To protect her self-confidence and to avoid feeling "bad" and "worthless" (words that unintentionally but often pop into her mind), she usually avoids taking part in exercises or training sessions where she is at risk of "revealing her shortcomings" or where other athletes might outperform her. Alexis thinks such scenarios would be shameful! Instead, she brings up minor injuries of hers and argues that she should be allowed to follow an individual training plan. Until now, she has remained confident that she will be selected for international competitions as the performance gap with other national team athletes has been rather big. She disapproves of the fact that younger athletes have joined the national team, and their rapid development triggers a feeling of insecurity about her position within the team. She has therefore started to pick on some of the younger ones, pointing out their flaws while avoiding training with them. Recently, the younger athletes have started getting some of the attention from the coaches that Alexis used to get when she was younger. To cope with the situation and relieve some of her stress, she turned to the other older athletes to discuss the situation and get support from them. Though initially the other athletes were less concerned about the situation than Alexis, she finds them to have been supportive, and the more she discussed things with them the more unanimous they have become regarding the inadequacy of the training offered them.

The younger athletes' group is, on the contrary, satisfied with the training and coaching they receive with the national team. They are highly grateful for the opportunity to join the team; they think it is awesome and see it as a great privilege because it will help them develop further as athletes. They push themselves to improve during training and are committed to getting the most out of each training session. They are aware of the older athletes' criticisms but view them as being a "bit whiny" as they don't share their dissatisfaction. As the older athletes appear to have a high status in the team, the younger ones are afraid to argue with them and don't dare to voice their views on the matter during team meetings. To avoid conflicts they remain silent when arguments arise. At the end of training sessions, the younger athletes immediately leave the training arena to avoid being influenced by the older ones' negative criticisms and attitudes. In the evenings, they sometimes get together for dinner and talk amongst themselves about how they view the team situation; they all agree that the older athletes' behaviors result in a negative atmosphere for the team and they often feel they are unfairly accused by the older group of things they have not done or are responsible for. The young athletes want simply to develop athletically and to get the most out of the training opportunities they now have access to as members of the national team.

During a training camp abroad, the situation between the two groups escalated with open conflicts. The head coach was approached by one of the young athletes who asked if something can be done about the situation in the team. The athlete thinks that the situation has escalated and the younger athletes on the team are feeling stressed and tense. The coach, who was unaware of the situation before training camp, approached the national federation body and asked for help from a sport psychologist consultant.

Thought Questions

Case Conceptualization

- What are the key aspects of this case and how did the various problems in the national team develop over time?
- Which psychological processes are involved at various levels (e.g., at the group level and the individual level)?
- Can you identify the primary psychological processes that presently maintain and increase the tense and agitated team climate?

Ethical Considerations

- Are there any potential "red flags" in the case that may be important to consider and where competencies in clinical psychology might be needed?
- How do you manage the information within and between different members of the team (e.g., what different athletes and coaches talk to you about)?

Diversity Awareness/Reflection

- What historical and contextual background information do you need to fully understand what is going on in this case?

- How would you get access to this information and from whom?
- How might political decisions made by the national sport federation impact the athletes, members of the national team, on a daily basis at an individual and group level?

Intervention Planning and Evaluation

- What would be your starting point with this case – do you start intervening at the group level, individual level, or both?
- What interventions would you consider and why?
- How do you involve and support the coach(-es) in your work with this case?
- How would you evaluate whether your work is helpful to the team?
- What evaluation criteria would you use and how would you assess progress?

5 Olympic Sport

"Being Mighty Warriors"

Shameema Yousuf

ZIMBABWE

Zimbabwe has traditionally been a strong sport nation producing talent now living as diaspora across the globe. Football is the most popular sport, with Zimbabwe Women's team, the Mighty Warriors, being present at Rio 2016. While Zimbabwe sport personalities are celebrated globally, in the history of the Olympics, their medal tally only stands at eight, seven of which were won by International Olympic Athlete Commission Chair, Kirsty Coventry in Swimming. The eighth medal was celebrated by the Zimbabwean Hockey Ladies team in their attendance at the 1980 Moscow (boycotted) Olympics, which saw 42 Zimbabwean athletes attend the games.

At the last games in Rio, Zimbabwe fielded 31 athletes – 9 male and 22 females who competed in seven different sports – making it the second largest attended in the history of Zimbabwean Olympic sport. In contrast to just seven athletes who attended London in 2012. The sports the athletes participated in at Rio were Women's Football, Rowing (Men and Women's Single Sculls), Equestrian (Individual), Shooting (Men's Double Trap), Archery (Men's individual), Swimming (Men's freestyle, Women's Backstroke), and Athletics (Marathon). In the past, Zimbabwe has also entered Alpine Skiing, Boxing, Cycling, Diving, Judo, Sailing, Tennis, Triathlon, and Weightlifting. Rugby 7s and Golf, both Olympic sports, may be future featured sports in Zimbabwe.

There is a social disparity in Zimbabwe in a nation where organized sport is mostly harnessed through the school curriculum. Only a small minority afford club membership and extra-curricular activity. Traditionally pre-independence sport was prioritized amongst the White population and has yet to be equally accessible to Blacks in the more regional and rural locations of Zimbabwe. Funding for the development of programs in more regional and rural locations is required, and this is operationalized by National Sport Federations (NSFs).

Talent is developed through the NSF's. Athletes develop recreationally, semi-professionally, or professionally through the NSFs. The federations provide coach education, guidelines, operational systems, and organize sport competition events to support youth and elite development in sport. With some funding that NSFs receive from the Zimbabwean Olympic

DOI: 10.4324/9781003089629-6

Committee (ZOC) and through other funding pathways, they can take their programs out to more local areas, towns, and cities throughout the nation. Emerging talents rise from low socioeconomic backgrounds alongside those from more privileged socioeconomic status, supported by the NSF coaches.

In Zimbabwe, the underpinning purpose of programs is an important consideration. For example, Zimbabwe's Health Minister Kirsty Coventry, who as mentioned earlier, is a seven times Olympic medalist and International Olympic Committee Athlete Commission Chair, has introduced many programs that encourage the participation of children in rural areas in swimming. However, these programs were initially targeted to improve African Blacks' mortality rate as a result of drownings. That is, swimming is a basic survival need, though many may develop further interest through access to school scholarships where swimming is part of the curriculum. Lastly, Zimbabweans participating in sport are a mix of cultural representation that includes diverse racial identities with wide disparities in socioeconomic status. They represent diverse religious beliefs, including traditional spiritual healing, mixed-gender, and sexuality. The youngest Olympic squad athlete is of age 17, and the oldest is 31.

Zimbabwean elite sport remains hampered by a lack of sufficient funding in a nation that relies on agricultural and mineral exploration, plagued by drought, poor economic conditions, hyperinflation, disease, and virus outbreaks, which have impacted industries across the nation; this includes sport. The ZOC is also a small national Olympic body, and many Olympic athletes do not have access to the full-time support of sport science and medical services, let alone psychological support. In more recent historical times of Zimbabwean sport, the medical and sport sciences systems have been fragmented with protocols and frameworks that incorporate testing and inconsistent support. However, after the appointment of a new head of medical services in 2017, there has been a shift to try and educate staff and implement changes to improve systems for the athletes and staff.

Funding remains the greatest barrier to elite success. However, some individual athletes can apply for the Olympic Solidarity Scholarship that is offered by the International Olympic Committee to National Olympic Committees for their candidates. For the athletes awarded just as Kirsty Coventry was, this will mean they can pay to attend more events and have some extra sport science and medical care. Indeed, Kirsty Coventry believes she would not have been as successful in her career without this scholarship while also on scholarship in the USA. Others rely on scholarships to universities abroad wherein their training environments they have access to resources and facilities that support their development as an athlete and person. Often, as an international athlete in a foreign land, they are motivated not only by competing at a high level in foreign competitive sport environments but also by the resources and opportunities available to them. The athletes who qualify for Olympic representation receive some sponsorship and support of the ZOC, having developed either internationally or through the NSFs.

The Zimbabwean sport federations and Olympic system have, in the past, had some limited exposure to sport psychology and hence some basic mental skills understanding. However, it has not been operationalized systemically on a regular basis. Given the structures and several pathways to developing as an elite Zimbabwean athlete and a diverse Olympic athlete body, psychological workshop delivery as commissioned by the ZOC were operationalized with three objectives:

1. That coaches gain insight into personal development, mental resilience, and well-being support in performance environments.
2. That coaches think about how they can support athletes in various stages of athlete development and transitions.
3. That coaches have a better understanding of working with female athletes and risk factors to well being and mental health given the recent success of two female sports teams.

With some funding for education provided through partner relationships of ZOC, educators and consultants are engaged to deliver topics that support coaches in the technical, tactical, physical, and psych-social elements of sport. Zimbabwe's partnership with the Norwegian Olympic Committee enabled funding to be set aside to deliver sport psychology workshops to educate coaches from the NSFs.

Thought Questions

Case Conceptualization

- You are running workshops for coaches whose athletes are on the elite Olympic/pro sport pathway – what would you consider given that many of the athletes will need to travel to national training centers, and some may move abroad?
- The coaches have never had much psychology education. What recent topics might you consider introducing to consider the environment, culture, and the importance of personal development?

Ethical Considerations

- In this case, the Olympic Committee has engaged the psychologist and will pay the bill for the first set of workshops. The delivery education for invited NSF coaches. Who is the client?
- What would you need to consider for equal access in the workshop delivery?

- Aware of the Zimbabwean utility outages, what are some of the ethical considerations in delivery? How would you contemplate utility failures?

Diversity Awareness/Reflection

- What contextual knowledge of the national sport culture do you have to help you in this delivery setting?
- What are some of the key considerations for delivering in an African country whose national languages are English, Shona, and Ndebele?
- How would you consider discussions with coaches whose team have diverse cultural representation?
- What knowledge may you wish to have about cultural practices in mental health and spirituality?

Intervention Planning and Evaluation

- In a country with limited telecommunication access that relies on generators for electricity when the grid is down, what are some of the key considerations and clarifications for the delivery of workshops?
- How would you engage coaches with education content?
- The Olympic Committee must evaluate the effectiveness of the educational workshops. How would you evaluate your effectiveness?

6 Paralympic Sport

"Classification Troubles in Wheelchair Rugby"

Grzegorz Więcław

POLAND

Ian is a 28-year-old international wheelchair rugby player with a spinal cord injury (SCI) at the level of sixth cervical vertebrae (i.e., neck). Such a disability somewhat affects his functional movements from chest down. He has no feeling in his legs, limited use of the trunk and good, but not full control of his shoulders, arms, and hands. He acquired his disability by diving into unknown, shallow waters at the age of 23. Despite an initial poor prognosis on his recovery, he made steady progress in rehabilitation and regained some additional function in his upper body making him a fully independent individual in daily activities. During the rehab period he sampled a variety of wheelchair sports and absolutely fell in love with wheelchair rugby.

Wheelchair rugby (originally called "murderball") is a fast-paced, full-contact team sport featured at the Paralympic Games since Atlanta 1996. It was originally created for people with SCI at the level of cervical vertebrae, whose severe disability made it too challenging to compete in other wheelchair events such as wheelchair basketball. Currently, the International Wheelchair Rugby Federation (IWRF) allows people with other disabilities to participate as well. A rule of a thumb in wheelchair rugby classification is that an eligible athlete needs to have some loss of function in at least three limbs. Thus, the modern game can be played by people with amputations, cerebral palsy, muscular dystrophy, and other medical conditions and physical disabilities, innate or acquired (Altmann et al., 2015).

Classification has been part of disability sports since the 1940s and serves the purpose of making the competition as fair and equitable as possible. In accordance with the International Paralympic Committee's (IPC) classification guidelines (IPC, 2015), the IWRF considers three key questions in the process of athlete evaluation: (1) Does the athlete have an eligible impairment for this sport? (2) Does the athlete's eligible impairment meet the minimum disability criteria of this sport? and (3) Which sport class describes the athlete's activity limitation most accurately?

In wheelchair rugby, eligible athletes are classified according to their functional abilities based on a point value ranking from 0.5 (lowest functional level) to 3.5 (highest functional level). With four players allowed

DOI: 10.4324/9781003089629-7

on the court, their maximum total point count cannot exceed 8.0 points. To determine a class in wheelchair rugby, an athlete undergoes a series of assessment tests focusing on the function of their trunk and upper body, especially the hands. The classification takes place prior to IWRF-sanctioned, that is, official events and is followed by observations during a competitive match (Altmann et al., 2015).

Despite how extensive the IWRF's Classification Manual (Altmann et al., 2015) is, some athletes are difficult to classify. One reason may be the atypical nature of their disability when confronted with a manual that was originally created for people with SCI. The other reason may be that they present characteristics of two different classes or fall in between classes leaving room for interpretation. Occasionally, such situations happen in real life and Ian's case is an example of that. It is also important to note that in wheelchair rugby, an athlete gets a permanent international classification (P) if they are assigned the same class three times in a row by three different classification panels.

Out of rehab, Ian quickly adapted to life in a wheelchair. Getting involved in sport definitely helped him regain more control over his life. He joined a local wheelchair rugby club and proved to be a fast learner of the game. On his first national classification he was evaluated as a 2.0 player, typical for tetraplegics with his type of injury. Wheelchair rugby athletes in this class have strong and stable shoulders that allow good pushing speed on the court. They can hold the ball firmly and chest pass with control from medium range, but lack proper hand function. They may be ball handlers on the court, but in most teams are not major playmakers (Altmann et al., 2015).

Ian was happy with his first classification and kept working hard on his individual development. He absolutely loved the sport and people he met through his engagement in wheelchair rugby. Slow but steady, training became the most important part of his weekly schedule. At some point he realized that he was no longer playing wheelchair rugby just for fun or as a rehab activity, but had become a wheelchair rugby player – a competitor.

At the age of 25, he was called to play with the national team and participated in his first IWRF-sanctioned tournament. His national class of 2.0 was confirmed by the international panel. The more he played, the more this 2.0 classification became a strong part of his wheelchair rugby identity; "I am a 2.0 player," Ian would say to himself. At the age of 26, he climbed higher and played at his first European Championship. Despite poor team performance, he stood out as one of the most prospective and promising players on the squad. His 2.0 class was confirmed again. Ian was very happy with the place he was in his sport and life at that time.

The following year, Ian did not compete internationally because the national team did not qualify for any major tournaments. It did not bother him much; he had his individual goals to achieve, mostly focusing on improving his general physical fitness and ball control skills. He was optimistic and very determined to return to the big competition stage as prepared as ever. When the national team made it to the next European Championship, hopes were high for Ian. He knew he had worked really hard the entire year improving his speed, agility, and endurance. He also had improved in some technical aspects, like pushing his body to have better control over the wheelchair and the ball. All in all, he felt ready and pumped up. He wanted to play a major role on the squad and help the team to achieve its goals, and everyone on the coaching staff was convinced that Ian could do as he intended.

Ian also knew that this pre-tournament classification would likely be the last one before he would go permanent in the 2.0 class. He was confident it was purely a formality; after all, he had gotten his 2.0 class twice in a row, without a blink. "A typical tetraplegic 2.0," classifiers had previously told him. He knew exactly what to expect from the classification process, having been in classification panels a few times in the past both nationally and internationally. While he was not overly worried about it, he was a little nervous – as much as anyone would be when facing any assessment whose outcome mattered.

On the day of the classification, he joined the panel of classifiers accompanied by the national team physiotherapist (a standard procedure in the wheelchair rugby classification process). The process started as usual with an interview and the review of medical papers, followed by the functional assessment tests. As the process continued, Ian began to feel that this classification appeared to be different from the previous ones; far more detailed than he remembered and definitely took a lot more time. The panel focused primarily on one of his hands claiming that current functional assessment test results were different from those two and three years ago. The classifiers' approach made Ian feel very uncomfortable; as if he had regained some function and/or he had cheated in front of previous panels. The tests went on and on. The classifiers debated the case and started doubting Ian's pure intentions. At the end of the session, the outcome was not favorable. To Ian's utter dismay, the panel labeled him as non-cooperative (NC). In the Paralympic sport classification process, it is the responsibility of the athlete to cooperate with the panel and give his/her full effort during functional tests disclosing maximum physical ability. Lack of compliance and/or attempt to conceal real capabilities during classification is treated pretty much like doping by IPC (2015), and the IWRF may sustain penalties over such athletes. Penalties vary in severity of sanctions and may include, for example, international class suspension and periodic or even permanent disqualification (Altmann et al., 2015).

Upon hearing the decision, it is no wonder that Ian was absolutely shattered. The NC class not only stopped him from playing in the upcoming tournament, but also labeled him publicly as a cheater and a fraud. He claimed that he cooperated fully and gave his maximal effort as Ian never had any intention to conceal anything from the panel. In his view, all he did was to keep training his disabled body to the maximum of its potential and this is where differences in assessment tests come from.

Ian was in tears when the head of the panel gave him another chance. Based on additional tests the panel classified him this time as a 2.5 player, which allowed him to compete with a new class in the upcoming tournament. Ian was frustrated with this decision as well. After all, he was a typical 2.0 wheelchair rugby athlete with SCI. He was convinced he could be an outstanding 2.0 player, but a class higher would make him mediocre at most. He certainly did not feel he could compete against other 2.5 players. His initial sadness and despair upon receiving the NC and then 2.5 turned into anger. Despite the negative emotions and having had his sense of player identity severely shaken, Ian did not lose control of himself, instead he swallowed this sour pill in silence.

The team manager appealed the panel's 2.5 decision; a common practice in Paralympic sports when a decision does not favor the athlete (IPC, 2015). However, after game observation and follow-up classification by a different panel, Ian got NC class again. Subsequently, he got disqualified from playing wheelchair rugby for at least six months. As it turned out, in the follow-up classification Ian desperately tried his best to fit with the 2.0 criteria. His attempt to do so was revealed, and the second classification panel took it as a clear indication to conceal functionality from his end when making their final decision.

Ian remained at the European Championship event with the team but had to watch the games from a distance. He was angry, disappointed, and felt that he was treated unjustly. He appeared to withdraw from the team's social life and allegedly turned to alcohol during the tournament.

As a sport psychologist on the coaching staff of this national team, you have known Ian for two years. Through your work with the team, you managed to build rapport with Ian, and you feel there is a level of trust to address difficult issues.

Thought Questions

Case Conceptualization

- What are the key aspects needed to best understand the meaning of Ian's class change and subsequent disqualification?
- How would you support Ian individually in facing his own adversity, keeping also in mind the interest of the entire team which is about to compete at the European Championship?

Ethical Considerations

- How would you deal with Ian disclosing sensitive information with regards to the classification process, often including personal, medical, and functional data?

- What potential "red flags" come to mind when thinking of various coping mechanisms in such situations?
- How would you collaborate with the head coach and other coaching staff members in order to provide the best support for Ian and the team?

Diversity Awareness/Reflection

- How would you broaden your understanding and perspective on the classification process and its implication on sport-specific identities of wheelchair rugby athletes?

Intervention Planning and Evaluation

- How would you approach supporting Ian as an athlete and as a person given one full year as a timeframe?
- What kind of psychological tools would you include when planning potential interventions with Ian?
- How would you best prepare yourself for different scenarios in this case?

References

Altmann, V., Hart, A., Newell, E., & Ungerer, G. (2015). *IWRF Classification Manual* (3rd ed.) (J. Erasmus, Ed.). International Wheelchair Rugby Federation. http://iwrf.com/resources/iwrf_docs/IWRF_Classification_Manual_3rd_Edition_rev-2015_(English).pdf

IPC Athlete Classification Code: Rules, Policies and Procedures for Athlete Classification. (2015). Bonn: International Paralympic Committee. https://www.paralympic.org/sites/default/files/document/150813212311788_Classification Code_1.pdf

7 Professional Sport

"The Outsider"

Charlie Maher

United States

This United States-located professional basketball organization has contended for the league championship on and off during the past decade. With regard to its basketball operations division, the organization includes a general manager, two assistant general managers, a head coach, five assistant coaches, a three-member athletic training staff, and two strength and conditioning coaches. The organization has employed these individuals for the past two to three years. The organization also had access to two licensed mental health counselors who were associated with a nearby mental health clinic. These counselors were available to players and staff for mental health concerns on an as needed and confidential basis.

During the recently completed season, sport psychology services were provided by someone who occasionally gave mental skills presentations and met periodically with players. He was not integrated into the basketball operations structure.

Once that season was completed, the consulting contract of that individual was not renewed by the organization. Consequently, the general manager, in conjunction with the head coach and other basketball operations personnel, formed a search committee. They decided to hire a sport psychology professional who would reside in the area and be employed as a full-time staff member.

This committee reviewed the credentials of a range of individuals who responded to a full-time mental skills coach's notice. After interviews and deliberations, the committee settled on their choice.

The individual that was hired – who will go by the name of Jan for this case study to protect anonymity – agreed to move to the area and be a full-time employee of the organization. He agreed that he would be referred to as the team's mental skills coach. He agreed to a two-year contract with the organization.

Jan possessed a bachelor's degree in psychology, a master's degree in sport psychology, and claimed to be working on becoming licensed as a professional counselor. His prior experiences in sport were with another professional basketball team as an assistant consultant. He also served as a mental skills consultant to a professional baseball team as well as being a mental skills coach for a college football team.

DOI: 10.4324/9781003089629-8

Jan began his work in July by meeting with the general manager, head coach, and other basketball operations staff members to discuss how he would function and provide services to the team. He proposed that his overarching goal would be to assist players in developing mental skills to be the best performers and people possible. His stated intention was to establish productive working relationships with players and with the coaches.

To initiate his work with the players, Jan held an orientation meeting with the team at the beginning of training camp (October). All of the players and coaches attended this meeting. Jan described his sport psychology background and then discussed what his role and responsibilities would be with the players and team. He stated that his role was going to be to assist them in developing mental skills so that they can perform well on the court and so that they can develop themselves as a person, over and above their profession of basketball. As part of the meeting, he handed out a one-page description of his roles and responsibilities.

The team consisted of 12 players (there also was a G league team of 10 players, but Jan was not asked to be involved with them during his first year).

There was considerable cultural, ethnic, professional basketball experience, and players' educational diversity on the team. Eight (8) of the players were African American and had attended three years of college or less. One (1) player, who was White, had graduated from college on the five-year plan. Two (2) players were of European descent, both of whom spoke English; they all had played professionally for several years in Croatia, Russia, and France, before playing in the United States. One (1) player was of Asian descent, played professionally previously in China, and spoke some English.

Jan began his work with the players during training camp (October). His intention was to meet with each player; establish rapport with each one and determine what mental skills on which they wanted to work. These introductory meetings seemed to proceed well; players were willing to talk with Jan about themselves and their mental skills needs. However, several of the players were resistant to having regularly scheduled meetings with Jan.

Through the first part of the season (November–January), Jan endeavored to meet, one-on-one, with each player to continue getting to know them and attempt to pinpoint mental skills for their collaboration. In this regard, Jan would attend practices and games, both home and away games.

During this time (November–January), Jan also sought to meet with the head coach and assistant coaches before and following practices. These meetings were informal, not scheduled, and very inconsistent. During these meetings, some of the coaches, as well as the general manager, asked Jan to tell them what the players were "deficient at mentally" and what Jan was talking about with the players. The coaches were interested in learning what Jan was working on with the players. However, they did not seem interested in being involved other than in terms of the players' mental games. A few of the coaches also told Jan what he should work on with the players, such as being more aggressive on the court and developing a better work ethic.

In addition, Jan sought to establish working relationships with the performance team, which consisted of the athletic trainers and strength and conditioning coaches. These relationships developed slowly but seemingly in a professional manner. Jan sensed that these performance team members were interested in the mental side of basketball. However, Jan sensed that he was not perceived as being part of the performance team.

During the second part of the season (February–April), the team was not performing as expected, based on their record, and it appeared that the team would not qualify for the league playoffs.

It was during the second half of the season that Jan experienced things "speeding up" for him concerning (a) fulfillment of his role; (b) responding in a helpful manner to players' needs and requests; (c) dealing with challenges provided by the general manager and the coaches; and (d) not having contact with the mental health counselors.

With regard to his role, Jan sensed that players and support staff perceived him as being someone who was there to talk to them, typically in an informal way. Some of the players referred to him as the team sport psychologist, others as the "mental skills guy," and one as the "team shrink." The athletic trainers and strength coaches started to recognize him as a mental performance coach, although they did not seek him out as an integral part of the performance team.

Jan was aware that there were differences in what he was being referred to by players and others. However, he did not think that this was any cause for concern at the time.

In terms of his interactions with players and his intention of helping them with the mental side of their basketball games, Jan had periodic mental skills sessions with seven of the players. His work with those players typically involved periodic meetings and interactions with them, either before or after practices, and sometimes at the hotel when the team was on the road.

The areas and goals in which Jan was involved in mental skills work with these players included: (a) balancing basketball with the rest of their lives; (b) maintaining self-motivation; (c) focusing on what matters in the competitive moment; (d) committing to a one play at a time mindset; (e) relating appropriately with family members and significant others, and (f) communication with the coaches about playing time.

With regard to mental skills work, Jan would meet with a player and discuss where they were concerning their mental goal. In addition, he provided some of them with advice, forms, and checklists based on the interest of a player in receiving and using that kind of material.

The players who were not meeting with Jan regularly, still had contact with him. Typically, they would talk briefly on the floor with Jan during breaks in practice, and he would provide them with words of encouragement as well as with his observations of their on the court performance. As the season progressed, the players seemed to be satisfied with Jan's contacts with them; and about seven of them continued to meet with him, although not always regularly.

With regard to the head coach and the assistant coaches, Jan worked at establishing a working relationship with them. However, it was with the assistant coaches that Jan experienced conflict and professional concern. Although the coaches allowed him at practices, he sensed that he was not providing them with useful guidance about the players. In this regard, the coaches would often ask Jan what he was talking about with the players when he met with them. These requests made Jan feel uncomfortable. On the one hand, he wanted to respond to the coaches' questions, while, on the other, he also was trying to maintain confidentiality.

In order to deal with this dilemma, Jan had a meeting with the head coach about the matters of confidentiality and what he could or could not relay to the head coach and other coaches. The head coach seemed to understand Jan's explanation about the confidential nature of player discussions. However, the head coach indicated that he wanted Jan to help the coaches learn about and understand the players from a mental and emotional perspective.

With regard to performance team members – athletic trainers and strength coaches – Jan believed that he was developing a collegial relationship with them. Over time, Jan was involved in general discussions with these individuals about matters such as mindfulness and composure. However, they never referred to Jan as a member of the team's player performance staff.

The area of mental health of players became a concern as the season progressed. Although he claimed that he was taking courses to become a licensed professional counselor, Jan was not a licensed mental health professional. Within this context, the licensed mental health counselors who were contracted by the organization to provide the mental and emotional support of players had not developed a working relationship with Jan. One of these professionals told the general manager that Jan was not in contact with him regarding player mental health referrals. Moreover, this individual also relayed to the general manager that he heard that Jan was overstepping his bounds from mental skills work to the area of mental health by providing advice to players with regard to their mental and emotional health.

As the season ended in April (the team did not make the playoffs), Jan felt that things were speeding up on him. In addition, when Jan asked the general manager about next season and his role, that individual was vague and seemed distant in his response.

During May, Jan contacted you to provide him with some guidance and direction for how to improve his work with the team and organization.

Case Conceptualization

- What information do you need prior to beginning work with this situation to enhance your effectiveness?
- What would you want to know more about with regard to the education and training of the mental skills coach?

Ethical Considerations

- Do the facts in the case suggest any "red flags"? That is, do you need to consider clinical referral, limits of your consulting capacity, competing stakeholders, etc.
- What concerns exist with how the mental skills coach related to the mental health counselors?

Diversity Awareness/Reflection

- What are your initial thoughts regarding the culture in this case?
- What is your opinion about how matters of ethnicity and race were addressed in the organization? What would you like to know more about with regard to these matters?

Intervention Planning and Evaluation

- Identify the potential issues that may be affecting the performer (mental skills coach)?
- What is your opinion about how the mental skills coach went about establishing productive working relationships with players?

Section 2

Cases from High-Risk Occupations

8 Police/Fire

"Back into the Breach"

Damien Stewart

AUSTRALIA

Ever since James could remember, he wanted to be a Police Officer. James' mother remembers him beginning to speak about being a Police Officer around the age of 10 or 11, however James has no memory of ever wanting to be anything else. James remembers sitting and watching cop shows and movies on television with his father, dreaming of one day wearing the blue uniform of the Police Force. James was also particularly interested in television shows that depicted the work of Special Weapons and Tactics (S.W.A.T.) officers, and hoped to one day try out for S.W.A.T.

When James first sought help, he was 33 years old and had been married to his wife, Kasia, for 11 years. James and Kasia had three young boys aged nine, seven, and four. James reported that he had an excellent marriage, that he was close with his three sons, and that there were no particular issues within his marriage, home life, or with any of his children.

At the time of this first meeting, James had been a Police Officer for a little over ten years. He explained that getting onto the Police Force was not as easy as he thought it would be and, like many young people who start down the road of becoming a Police Officer, he had to apply several times before he was finally accepted. He was finally accepted onto the Police Force on his third attempt.

James was a little coy as to why he had failed the Police Force application process the first two times; however, he finally revealed that the recruiters felt he was too immature at the time and encouraged him to go back out into the world to get a little more life experience before applying again. James eventually went further to reveal that he had difficulty with some of the aptitude testing that was part of the application process. However, James discussed that he had hired a tutor and went to work on improving his skills before he applied for the Police Force the third time. Once he felt confident, James applied to the Police Force again, passed all of the entry examinations and requirements, and was invited to join the Police Force. James had to wait about three months to begin the intense training at the Police Academy, but he was in and he was excited.

Once he began at the Police Academy, James found the process to be a breeze. He struggled slightly with some of the legal studies, policy, and procedure learning, but had no problem with the physical training such

DOI: 10.4324/9781003089629-10

as physical training, weapons and tactics, and high-speed driver training. Ultimately, James graduated from the Police Academy with flying colors; he didn't "dux" the course (meaning he did not finish top of the class), but he was closer to the top of his class than the bottom.

Once he graduated from the Academy, James, like all new Police recruits, was assigned to a Police Station where he spent time in uniform performing a General Duties Policing role, patrolling the streets of the community. General Duties Policing is the most dangerous and important of all Policing roles, yet it is generally staffed by the most junior and inexperienced Police while they hone their skills and work toward a more specialized role later in their career. James was sent to a large regional community approximately 100 kilometers north of the state's capital city where he worked in uniform for six years. Toward the end of his sixth year, James was accepted into a plain-clothes Policing role where, along with his team, he conducted targeted policing of trouble spots within the community.

Having put in his time in uniform, James had not lost his dream of becoming a S.W.A.T. Officer and patiently waited for the time that he could apply to join this elite unit. Not only was James up against a requirement for solid Police experience before applying, positions within S.W.A.T. did not come up very often, so he found himself waiting for the moment when he could apply. Finally, the opportunity arrived during his eighth year of Policing.

The process for becoming a S.W.A.T. Officer is not for the fainthearted. Those wanting to join S.W.A.T. must first go through a vigorous selection process before beginning the specialized training needed for the role. First, recruits must complete a battery of aptitude and psychological testing, an interview with senior officers, and pass a rigorous fitness test. Once a potential recruit passes all of these tests and is deemed suitable to continue the selection process, they are invited to begin the selection course. The selection course requirements are based on similar selection courses for military special forces programs, although not as long or as intense. The S.W.A.T. selection course lasts for three days and is designed to test a potential recruit for vital abilities such as strength, endurance, ability to operate under pressure, leadership, teamwork, and ability to follow direction. If a potential recruit passes this phase of the course, they must then earn a positive subjective assessment from those S.W.A.T. members running the course; essentially, determining whether or not they feel they could work with the potential recruit. If one member declares a dislike for the recruit, he/she will not be selected to S.W.A.T. Even if selected, potential recruits face months of training before they are declared competent Operators and allowed to be deployed on S.W.A.T. operations.

As the intake session progressed, it soon became clear that James was not simply looking for assistance in becoming a S.W.A.T. Operator, but in helping him through the process a *second* time. James declared that 12 months prior, he had gone through the entire selection process and passed.

However, once he began the training phase after having successfully completed the three-day selection program, he returned home for the weekend to see his wife and children. Upon getting ready to drive back to the training facility on Sunday evening, ready for an early Monday morning start on the shooting range, James' middle child came to him sobbing terribly, and begging him not to leave again. James sat and spoke to his son with his wife and eventually managed to calm his son down so that he could leave, but the experience really got to James, "It broke my heart." James could not stop thinking about his son on the long drive back the training facility and on Monday morning, after not having gotten much sleep, he informed the training instructors that he would need to pull out of the course as he felt he could not be away from his family for extended periods of time. Although surprised, the training instructors understood and were ultimately grateful James had been open and upfront about his dilemma. They informed him that if he ever wanted to try again, he would be most welcome.

Now, some 12 months later, James has decided to try again. In applying to join S.W.A.T. for a second time, they were happy to accept the results of all but one of his assessment items from the first time he applied: he would have to pass the three-day selection course again. The selection course was scheduled to run exactly 14 weeks from this first consultation with James.

In considering another attempt at the selection course, James reflected that the first course was a "mental health trauma." He described how, for three days, he and his fellow applicants had been put through their paces that included a 20-kilometer pack march, a 10-kilometer run, teamwork drills that included carrying a 100-kilogram dummy on a stretcher up a steep hill to the summit, navigation drills, and swimming across a wide flowing river in full fatigues, while being sleep deprived, low on caloric and water intake, constantly being yelled at, and having every decision second-guessed.

James admitted that although he passed the course last time, the trauma of what he had gone through was playing on his mind, feeling as though the course had "fried me last year." He described feeling trepidation at trying again and that he had demons in his head about what he went through and expected to go through again. James expressed that he was scared of failing.

James discussed issues that he felt were leading to his lack of confidence. First, James outlined that he was not training anywhere near as hard as he did the first time around. Second, James had told a few people at work that he was trying again and all the nay-sayers had begun to come out of the woodwork. Like any opportunity in life that only a few people can genuinely take hold of, it became clear that James was experiencing jealousy from his colleagues who perhaps wished they could become S.W.A.T. Officers but couldn't for a variety of reasons, such as inability or family circumstances.

At the end of the first consultation, the intake session, you and James determined that you could work together and skimmed over some things that might assist James in getting ready for the selection course. You discussed some initial strategies James could implement straight away and scheduled to meet again in a week.

As with all Police clients, you asked James to complete two psychological inventories prior to the first consultation: the Depression, Anxiety, and Stress Scale 21 (DASS21) and the Post-Traumatic Stress Disorder Checklist (PCL). On the DASS21, James scored Normal for Depression, Normal for Anxiety, and Normal for Stress. On the PCL, James scored 19/85 – a score well below clinical concern.

Thought Questions

Case Conceptualization

- What information do you need prior to beginning to work with this individual in order to enhance your effectiveness?
- What specific goals does it seem James wants achieve from his consultation?
- What specific concerns does James have that will need to be addressed in your work with him?

Ethical Considerations

- Do the facts in this case suggest any "red flags?" That is, do you notice the need to consider clinical referral, limits of your consulting competency, competing stakeholders, etc.?
- Do you currently have any responsibility to disclose to the Police Force that you are working with James? Why?
- Remembering that James is moving into a high-risk role. If the Police Force learned you were consulting with James and requested a copy of James' case notes, how would you respond?

Diversity Awareness/Reflection

- What are your initial thoughts regarding the culture involved in this case? These may be related to the performance arena or the individual performer.
- What sort of specific cultural considerations would be important when working with Police Officers?

Intervention Planning and Evaluation

- Identify the potential issues that may be affecting the performer (these may be personal and/or performance related). From your list, prioritize the top three issues that you believe are having the most immediate impact.

- What issues is James facing in preparing to enter the selection course, both from a non-performance and performance perspective?
- What are some non-performance related issues that might need to be addressed? Do you feel competent to work with these issues?
- What barriers might James face in preparing for the selection course?
- Are you familiar with the inventories used during James' intake? Do you know how to use them effectively or what the scores might mean? How would you respond to the scores if they returned a score that was of concern, especially due to James moving into a high-risk role?

9 Military/Special Warfare

"Life Lessons and Leadership"

Elizabeth Beaubrook

United States

It is the third straight day of training out in a remote training area in the swamp of Louisiana. The unit you're working with is in the middle of a train-up for a deployment cycle to the Middle East. As the mental performance coach assigned to the battalion, you work closely with all of the different companies, each with their own specific job critical to the mission ahead. The group you are with now is a platoon of combat engineers you have been with non-stop for several months as they navigate the necessary training to ensure they are prepared and skilled to deploy. Today, you are observing training as they continue battle drills focused on urban combat. As they enter and clear rooms, you are in awe of how real the training seems. This installation, like many others, has makeshift villages set up with simulated enemies—a mock-up representation of the villages these soldiers will be seeing in Afghanistan in a few short months. It is a training exercise, but it is *intense.*

As you watch the battle drills unfold, it seems like there is a constant and repetitious battle rhythm marked by yelling coming from the squad leaders—"BANG, BANG, BANG! CONTACT," "STOP," "DO IT AGAIN." It is clear to you that the squads are struggling with the objective of the exercise. You cannot be totally sure of the reason, but to your observer's eye, it seems like the lack of communication might be partially to blame. From a performance standpoint, lack of communication is a huge problem considering that successful battle drills are swift and coordinated movements of multiple four or five-man teams; an even bigger problem when you consider the dire consequences if they are not done just right. They *have* to get this right.

You stick around for the After-Action Review (AAR), which is a chance to evaluate the training effort. Unit members systematically review what was supposed to happen, what actually happened, and what was successful/unsuccessful with the effort, and determine what should be done differently in the next training event. It is a gold mine for information to inform your work with the unit. As you listen in, the soldiers echo what you observed: the communication was not good enough, especially when things got intense. You scribble some notes to use for planning the next session with the platoon and, as you close the notebook, you see one of the squad leaders, Staff Sergeant (SSG) Michaels, approaching you.

DOI: 10.4324/9781003089629-11

SSG Michaels is the soldier who originally invited you to come work with the unit—he believes in the importance of mental toughness and attributes his ongoing ten-year career in the military to his ability to develop that characteristic. SSG Michaels is a battle-hardened soldier, having been on numerous deployments to some of the most volatile combat zones. He is the kind of guy that lives and breathes this work and looks forward to the next deployment. He outwardly says he does not understand how you can teach mental toughness, but that he will fully support whatever you need if it helps his guys.

"What's going on, Michaels?" you ask as he approaches.

"Clearly not talking, that's for sure," he retorts. You chuckle a little, shaking your head in agreement. SSG Michaels is not one to add any fluff. "I need your help … I can't get through to Green and it's screwing us," he continues.

"Green …? What's up with him?"

"Look, he has so much potential. When we talk about this stuff back at the company area he can answer any question you throw at him. He's a high-speed soldier … until he's not. He freezes when we get out here! He can't make a decision to save his life—literally. And now that he's a team leader, not only will he get himself killed, but probably all of his buddies, too."

"He's having trouble making decisions during live training," you repeat back, making sure you're tracking what SSG Michaels is sharing.

"That, and he's not communicating. *He* should be the one yelling out there. I'm not going to be able to hold his hand in Afghanistan. I need him to step up and lead his guys out there."

SSG Michaels has a problem, indeed.

"What have you tried to address the issues?" you ask, assuming SSG Michaels wants advice on how to coach Private First Class (PFC) Green.

"Everything! I don't know what to do … I say it's your turn to try," he says, scoffing in pure frustration. "I'm sending him to your office next week—you tell me a time that works for you and he will be there."

You shake SSG Michaels' hand, knowing you have your own hands full with PFC Green.

Next week, PFC Green arrives early for your meeting. You welcome him into your office and he thanks you for making time to meet with him. He shares that he has been thinking about coming to see you since you started working with his unit and SSG Michaels gave him the push he needed. Since PFC Green mentioned SSG Michaels, you decide to ask why he thought SSG Michaels may have wanted him to come to meet with you. PFC Green replies "I'm not sure … I'm not a dirtbag soldier for him, but he's always on me about how I need to step up. He always yells things like, 'Say it like you mean it, Green!' Maybe it's a confidence thing in his eyes …"

You nod, considering what PFC Green shared and ask, "Well, what made *you* want to come see me?"

PFC Green tells you he's had a lot on his mind and just wanted to talk it through with someone else for some perspective. You notice he's speaking very generally, maybe a little reluctant to share what he has been facing most recently. You decide to hold off in asking about his present struggles for now and instead start with getting to know him personally a bit more.

"Well I can certainly be a listening ear and try to provide the perspective you're looking for. We'll get there, but I think it might be helpful for you to share a bit about who you are and the journey you've had leading up to this point," you tell him.

PFC Green nods and begins to tell you his background. He shares that he grew up in a small town in Colorado and joined the military because he was not quite sure what he wanted to do after high school. He toyed with the idea of becoming a police officer or EMT because he likes serving people, but decided maybe the military was the ultimate way to do that. Plus, joining the military would allow for him to "get out of the small town full of small minds." PFC Green shared that he felt like he did not always fit in growing up; he has a different way of thinking.

PFC Green said he enjoyed boot camp and the camaraderie he shared with the other basic trainees, some of whom he still keeps in contact with today even though they are stationed all over the world now. After basic training, he went to Advanced Individual Training (AIT) where he learned how to perform his job as a combat engineer. He excelled there, but said things were "different" than at basic training when it came to his social life. His AIT was at a small remote installation and he said it reminded him too much of what he was trying to escape back home. PFC Green stuck with it and was assigned to his first unit which is where he is now. PFC Green said that this installation was the last on his list of places he wanted to go, but he was trying to make the most of it.

Soon after getting to the unit, they deployed to the Middle East right at the end of Operation Enduring Freedom. PFC Green said it was great to get to apply his training down range, but it was a pretty low-key deployment all things considered. Soon after getting back home, several of his good friends moved to other installations and he was met with the reality that he is stuck in a small town environment with few friends.

PFC Green explained that things have been pretty boring at work as well. They have been in a train-up for another deployment for "what seems like forever" but the actual prospective dates keep getting pushed further and further back. He admits to having a hard time staying motivated to train day-in-and-day-out with no light at the end of the tunnel, especially when his first deployment did not necessarily live up to his expectations and the stories he has heard from SSG Michaels about his time down range. Plus, PFC Green shares that the other soldiers in his unit have been really frustrating him lately; it seems that people are not as "on point" as they should be. As a newly promoted team leader, it falls on PFC Green to keep his guys in line and performing to standard.

When PFC Green mentions his new role as team leader, you ask him to share how that transition is going. He reports that they have recently had some turnover in the unit which is part-and-parcel of military teams. But, as a result, PFC Green now finds himself in a leadership role. He says he thinks he is doing a decent job, but he is pretty nervous about the new role. He continues on to tell you that he knows he is young and generally quieter than other soldiers, but that SSG Michaels seems to believe in him. He tells you SSG Michaels "really pushes him" and recently nominated PFC Green to go to the Soldier of the Month Board where top soldiers compete against peers to prove their knowledge and skill to senior Non-Commissioned Officers.

To you, it seems like PFC Green has everything going for himself but you are surprised to see he is now visibly upset. You check in with him and tells you "I'm overwhelmed and nervous and have been thinking about a lot … the military, going to school, my guys … I'm really confused about where to go from here. I just want to believe what SSG Michaels keeps telling me—that I am capable. SSG Michaels is counting on me. I just know I'm not doing as well as I should be doing at work. I need help."

Thought Questions

Case Conceptualization

- What information do you need prior to beginning to work with PFC Green in order to enhance your effectiveness?

Ethical Considerations

- Do the facts in this case suggest any "red flags?" That is, do you notice the need to consider clinical referral, limits of your consulting competency, competing stakeholders, etc.?

Diversity Awareness/Reflection

- What are your initial thoughts regarding the culture involved in this case? Consider the military culture and the role of PFC Green's experiences growing up.

Intervention Planning and Evaluation

- Identify the potential issues that may be affecting the performer (these may be personal and/or performance-related). From your list, prioritize the top three issues that you believe are having the most immediate impact.

Additional Questions

- What elements of PFC Green's upbringing and past experience may be influencing his current situation?
- How does the PFC Green's relationship with SSG Michaels play a role in PFC Green's experience within his unit?
- What are some potential underlying beliefs or values that PFC Green has about leadership and performance based on his reaction to his new role and tasks at work?
- What are some considerations you have to be aware of as you navigate the "command-directed" referral from SSG Michaels for PFG Green to see you?
- What contextual knowledge do you need to know about the military to allow you to be a competent practitioner in this case?

10 Medical Professionals

"Mary Sue"

Doug Newburg

United States

Mary Sue is a 50-year-old grandmother. She loves to dance. In fact, she is a dance instructor who leads a dance class for men and women once a week. She works as an administrative assistant to one of the best surgeons in the surgery department of the university hospital in her hometown. Her coworkers view her as a leader that lacks the title. She is funny, kind of old school cool, and fun to be around without sacrificing her professional de meanor. She runs the office well, helps scheduling patients, and is a reliable link between the hospital and the doctor she works for directly.

On her way home from dance class one night, she feels a pain in her abdomen. She tries to ignore it, even when she gets home. Her stomach had bothered her before, but nothing that required serious medical interventions. This feels different. She feels different about it. She thinks about calling her daughter but decides to wait until the morning hoping it disappears. As she lays in bed that night, Mary Sue shifts around trying to find a position that alleviates the pain. In the background, her favorite country music plays on a loop, taking her back in her mind to the class she had led earlier that night. The class was fun, lively, flowing. She thrived on this class, on the dancing and the instruction she gave to new students. She had recruited several people from the hospital to take the class. They seemed to enjoy it which made her happy.

Her clothes for the next day hang on her closet door and she fell asleep smiling to the rhythm of her favorite music from the class. She would be fine, she told herself. Morning brings her usual routine of breakfast and coffee, a shower, and enjoying getting dressed in one of her favorite outfits. She liked looking good while also presenting the professional side of her at work. She liked being taken seriously at work without taking herself too seriously. She enjoyed her work, liked her boss, and felt that she played a critical role in the care of the patients her boss cared for. She spent as much time on the phone and in the office as anyone. She was the patients' introduction to everything, their guide through the system, and their go-to contact when things went wrong. She was one of the lifelines that made things work well. She knew it and embraced it. How she felt each day mattered because it mattered to the patients whose lives were in their hands.

DOI: 10.4324/9781003089629-12

Doctors, especially surgeons, and even more so heart surgeons, rely on their assistants (or whatever title is used for these critical roles). The power these people have in the system is often unrecognized. They can make or break the experience for the doctors or patients, which means they can also impact the nurses, technicians, etc. They often start the flow of information and can keep it moving, slow it down, or stall it altogether. The residents in a surgery practice also rely on people like Mary Sue. Surgery is a busy 365/24/7 experience for all hands on deck. Patients need to be where they need to be when they need to be there. Residents at an academic hospital need all the information processed and delivered in real time before they can prep their patients, before moving them to the pre-operation area. Anesthesiologists need their time with the patients. Forms must be checked, double-checked, signed, meaning someone is taking responsibility for every step of the way. Any delay, any absence of the needed information and authorizations, can have real-life consequences. Insurance information needs to be accurate and complete before the surgeon can put a scalpel to skin. Mary Sue is an essential link in this process.

At work the next day, she feels fine, as if nothing had happened the night before, so she dismisses mentioning it to her boss when he asks how the class went the night before. "It was great," she smiled, "like it always is." The doctor smiled back knowing how much it meant to her. As soon as he left for the operating room in the hospital across the street, she turned her music on, some early 50s country music. "The real stuff," she thought to herself.

The phone rang. Her daughter was calling to say hi which she did several times a week. Mary Sue considered telling her daughter about her stomach pain but decided not to.

"Mom, what's wrong?" her daughter asked.

"What do you mean?" Mary Sue asked. "Nothing. I'm fine."

"Moooom. I know that silence too well. Tell me what's going on," she demanded.

"Ok, you got me. Nothing big. I had some bad stomach pain last night after class," Mary Sue said. "I feel fine today."

"Promise me if it comes back, you'll see someone about it. You do work at a hospital, you know!"

"I promise."

They talked a little longer and the conversation unnerved her. The worry she felt the night before swept through her. The thought that maybe she'd say something about it to her boss crossed her mind again and then she buried herself in her work.

A week passed and it was time for class again, but as she was getting ready to leave her house the pain returned, but a stronger, stabbing pain that doubled her over. She was alone at home and tried to relax, laid down on the couch. The pain was so strong she forgot all about her class for the first time in her life until someone called asking her where she was. She tried to steady herself, to get up and go to class until she doubled over again and dropped the phone. Mary Sue gathered herself and told the person on the phone she couldn't make it tonight, that she wasn't feeling well. She asked the person if she could cover for her.

With the class covered Mary Sue headed into her room and collapsed on the bed. No music. She was scared. She thought about calling the hospital, using her connections to get to the right person, probably a surgical resident on call who was covering several services at once. Then, as so many people do, she thought about how she couldn't be sick, couldn't take time off, that things in her office would fall apart or be a mess when she got back. She couldn't be sick. She wouldn't allow it. Then she passed out or fell asleep depending on how one looks at it.

The next morning Mary Sue woke up to her daughter calling again. Mary Sue told her what happened, and her daughter demanded she get looked at that day. She threatened to tell Mary Sue's boss if she refused. Mary Sue knew she had no choice, so she agreed.

As she drove to work, Mary Sue forced herself against every instinct to call her boss, to find the right doctor to see her, and to make the arrangements as soon as possible. All she could think about was letting everyone down, how her boss and the patients and the patients' families needed her. She, and the others in the system like her, were the hubs of how work got done, how people made it seamlessly through the system—from parking passes to wheelchairs to insurance forms to x-rays, etc. If something went wrong, they were the fluid that made the workflow and worked out the kinks. Just like her dance class. They were the first people the patients' families would call. Most of the time Mary Sue liked how that felt. She didn't like letting people down.

"You need abdominal surgery." That's all Mary Sue heard the doctor say. She was scared. Routine surgery for the doctor, but rarely routine for a patient. Anesthesia. A scar left on her stomach. No matter how much one works at a hospital, it is different when you're the patient and everyone knows. No matter how many people one knows in the hospital, it is different when the doctors, the surgeons, are doing invasive procedures on you.

After the consultation with the doctor, Mary Sue was discharged to her daughter who drove her home. They were silent on the ride. The doctors had mitigated the problem enough that surgery could be scheduled for a month later. Mary Sue had time to hire a temporary replacement and train her as best she could until it was time. She had enough medicine to prevent further damage to her stomach and any pain, but these meds were not treatment and the surgery was not optional. The next morning, Mary Sue returned to work.

In the weeks ahead she had twice the workload; she had to handle the current patient rosters of clinics and surgery schedules as well as train her replacement. Her fear of letting people down returned. Maybe she could put off the surgery longer. She felt good, better even. Dance class was going well, and she rejoiced in the movement and music given how bad she was feeling before. She knew at the root of this she was scared—scared of the surgery and scared of not being there for the people who relied on her even if they didn't know it. Mary Sue was scared of at least two things. Like a lot of patients, and performers, she was more afraid of letting people down than she was of being hurt. She knew this. Given the freedom, also like a lot of people, she would choose to keep working rather than have the surgery. She wants someone to talk to, someone outside her circle of friends and family, who she fears letting down the most.

Thought Questions

Case Conceptualization

- Why a case about a patient instead of a medical professional?
- What does Mary Sue need from you? Why? How do you deliver that?
- What do you need from Mary Sue in order to help her?

Ethical Considerations

- What ethical considerations arise related to the hospital and healthcare requirements?
- How would you help Mary Sue without access to the hospital?
- How do you stay in your lane? How do you interface with the hospital and the staff? How do you make that decision?
- What role would family have?

Diversity Awareness/Reflection

- What contextual knowledge of the healthcare system do you have to help you in this setting?
- What contextual pieces are missing that you would like to know more about?

Intervention Planning and Evaluation

- How do you deal with family, specifically family who might be against the type of assistance you can provide?
- How do you deal with doctors who might be against this or causing a problem for Mary Sue in doing this work successfully?

Section 3

Cases from Performance Domains

11 Dance

"Choking on the Dance Floor"

Ashley Coker-Cranney

United States

Rayne is a beautiful young dancer in her early adolescence. With a sister also in dance, her whole family is heavily involved and very supportive of Rayne's hopes and dreams on the "marley" (i.e., dance floor). Rayne started dancing later than many, joining her first dance class at six years old. Rayne's parents never wanted their children to feel rushed or pressed, so they waited to enroll their daughters in any type of sport. The more friends who talked about their extracurricular activities, the more Rayne questioned her parents about when she would be "old enough" or "big enough" to participate as her friends did. After her parents had several discussions together, and with Rayne, they agreed that it was time to enroll her in her first season, and dance seemed to be a good choice to foster coordination, balance, flexibility, strength, musicality, fellowship, and followship.

Early on, Rayne was a bit uncoordinated and had to work hard to find the lines she needed to look the way her teacher did. Every now and again, her perseverance would pay off and she would catch a glimpse of the right line or angle before it faded away. When she came home from dance class, Rayne would prance about the house, showing off her undeveloped "relevés" (i.e., tiptoes), spotless "pirouettes" (i.e., spinning around wildly; spotting involves looking at a single place as the dancer turns to avoid falling over or losing count of how many revolutions she completed), and elementary "jetés" (i.e., leaps). By her parents' account, Rayne was in heaven.

Soon, she sought out online videos of dancers, teams, and performances whenever she could. She would imitate what she saw and relish in the attention she received from her parents and extended family as she "danced with the big kids." As time passed, Rayne asked her mother and father to get her to the studio earlier and earlier to see what the other dancers were doing while she mimicked them in the hallway. On occasion, they would notice her and applaud or wink or high five her for her efforts. Her voracious appetite for learning and trying new things on her own eventually earned her the notice of several teachers and the studio director. Offered an opportunity to compete as a soloist at seven, Rayne jumped at the opportunity. Her parents, although supportive, were more hesitant about

DOI: 10.4324/9781003089629-14

signing their daughter up for competitions at such a young age. Neither had a personal history with dance and they were unsure about how competitive dance would align with their family values (e.g., perseverance, passion, family focus, growth, vulnerability), affect their daughter's development, and impact their own personal experiences as dance parents. They often jokingly questioned whether the stories of "dance moms," and portrayals of dance parents on television, were accurate.

After speaking with parents of other young, competing dancers, and the studio director, Rayne's parents finally relented, saying, "What's the worst that could happen? We will handle what we need to." They surprised Rayne later that night, letting her know that she would have an extra day at dance the next week as she started learning her very own routine. Rayne was over the moon with excitement. She shrieked and jumped and hugged and went to bed, dreaming of who she might be on stage. At school, she told all of her friends how special she was to get her very own solo. Most were happy for her, many didn't understand why she was so excited to start with, and some began to tease her, mocking her excitement and taunting her "not to mess up in front of everyone."

Rayne attended her first rehearsal for her solo a few days later. Her unbridled excitement on the way to the studio turned into dread and fear as she began walking up the stairs to her lesson. Suddenly, Rayne's inner voice echoed the voices of her classmates. "Don't mess up." "Better not forget something." "You're not good enough for your own dance." "How are *you* going to be able to dance all alone in front of anyone?!" Rayne turned to run down the stairs, only to be caught in her mother's arms as she immediately began sobbing uncontrollably. As her mother pleaded with her to tell her what was wrong, Rayne explained that she just couldn't do it. She said she needed to find a bathroom—she was going to be sick. Her mother escorted her to the bathroom and stood outside to wait. Nothing happened, but Rayne couldn't shake the sick feeling in the pit of her stomach. She breathed and thought and cried, but nothing changed. With an urgent resolve, Rayne picked herself up and walked past her mother, into the studio to her waiting teacher. As her mother followed her to explain to the instructor what had happened, Rayne shot her a glare as if to say, "Don't you dare tell her what happened!" As the lesson went on, Rayne was determined not to show a sign of weakness or uncertainty, but she did notice a growing urge to gag when rehearsing big tricks or long combinations. It started small, as a short convulsion in her stomach. At first, it was just enough to take her slightly off balance. As the end of the hour grew nearer, Rayne requested several breaks mid-rehearsal to stop and catch her breath and suppress the rising need to gag—maybe vomit—as she danced. Rayne pushed through and sat with her teacher for a time to discuss how she did, what she wanted to do, what competition is like, and what to do next.

Thirty minutes after her lesson Rayne met her mother in the waiting room and walked in silence to the car. In the safety of her mother's vehicle,

Rayne let herself go, firing off everything she felt that evening. Her mother listened intently on the drive home, careful not to insert her own thoughts or feelings, but scrambling in her own head to understand what had happened, why, and formulate a plan to help her daughter. When they arrived at home, Rayne begged her mom not to tell her sister or father and promised she would "get it under control." Confused, but supportive, her mother agreed, but tucked the night away in her memory.

The next few weeks flew by and the need to gag, paired with the determination not to, became commonplace for Rayne. Rayne and her mother didn't discuss her first solo rehearsal for the rest of the season. Watching her daughter, Rayne's mother started learning the cues immediately prior to Rayne's gagging. Rayne had done an exquisite job hiding her urge to throw up, sometimes playing it off as intentional choreography to contract a bit harder or exhale a touch more forcefully. She was able to cover up her feelings for everyone around her, even, sometimes, herself. Rayne tried her best not to think about it, telling herself "I'll be fine," "It's nothing," and "I don't have time for this, get it together!"

Rayne's competition season came and went, and she scored well at each, often placing in the top five, if not the top three, soloists. Not a single judge seemed to know her secret, but Rayne, and her mom, knew the growing difficulty she was having. At the end of her first competitive season, Rayne attended a national competition, placing second, and reveled in the sense of accomplishment her tiaras, trophies, and ribbons won her.

The next several years blurred together as Rayne moved from one class to the next, one group to the next, one solo, duo, or trio practice to the next. She became better friends with some dancers and lost touch with others. She saw familiar faces at competition and began targeting rivals she wished to vanquish. At conventions (long weekend events where dancers can take classes with industry greats), she was singled out by choreographers, praised for her work ethic, daring, and challenge mindset. Starting slowly, but gaining steam, she was invited to apprentice at future conventions, where she would act as a young model on stage, demonstrating the combinations and pieces the choreographers taught to audiences of hundreds of other young dancers.

The familiar feeling of gagging, talking herself out of it, and trying to hide it from her audience came up many times, each time she took the stage. With time, Rayne learned to manage her gagging by restricting the type and quantity—if anything—of food she ate before practices, conventions, and competitions. She developed a strict list of the types of foods she could, and would, eat to minimize the frequency and intensity of her gagging. Some foods were banned based on taste and others for texture. Sometimes, Rayne would cut out foods because she ate them directly before a particularly difficult time controlling her gags, but other times, it was enough to fear they *could* make her gag. Ultimately, Rayne's greatest fear was not that she would gag—she *knew* she would. Instead, her greatest fear became that she would gag so badly on stage that she would vomit in front of an audience of friends, teachers, judges, or strangers. That fear drove nearly every decision Rayne made about her body and her sport. Unsure how else to help, her parents supported her efforts, seeing her picky pre-performance palate as "quirky" rather than concerning. In no time, Rayne took pride in her ability to control the gagging, but a growing fear that someone might notice and "out" her came to a head in middle school. Her nights of crying in private before a competition or convention were harder to hide from her parents, and she begged her mother to reach out to you for help so she might finally control her fear and anxiety.

Thought Questions

Case Conceptualization

- What information do you need prior to beginning to work with Rayne in order to enhance your effectiveness?
- What strategies would you use to gather that missing information?
- What is/are the most pressing concerns for this client? How do you make sense of the etiology of her challenges and the impact of her existing coping skills on performance?
- What is your prognosis for Rayne, at this point in time, without intervention?

Ethical Considerations

- Do the facts in this case suggest any "red flags?" That is, do you notice the need to consider clinical referral, limits of your consulting competency, competing stakeholders, etc.?
- How do you plan to navigate the client's status as a minor? Her parents' choices on how to handle her challenges?

Diversity Awareness/Reflection

- What are your initial thoughts regarding the culture involved in this case? These may be related to the dance arena or Rayne.
- How might your level of experience in/with the sport culture affect your work?
- How might your personal experiences in sport, and life in general, affect your work with this client? Her parents? How would you address your biases?

Intervention Planning/Evaluation

- Identify the potential issues that may be affecting Rayne (these may be personal and/or performance related). From your list, prioritize the top three issues that you believe are having the most immediate impact.

- How does your consulting philosophy affect your approach with this client? How (in)compatible is your theoretical and practical orientation to the needs of this client?
- What treatment goals will you use as your basis for evaluation? What methods would you use to evaluate progress? When would you use them?
- How would you address Rayne's current approach to managing her gagging?
- How would you decide when/where/how to address Rayne's peer relationships?
- How might you approach Rayne from a symptom-management versus cause-focused perspective?
- When/where/how would you enlist Rayne's parents in the process, given what you currently know about them?

12 Music

"The Returning Pop Singer"

Kaori Araki

JAPAN

Yoko is a 36-year-old female vocalist and a member of a popular band with two other male guitarists. The band made their major debut in Japan when she was 22 years old. After releasing over 30 single songs and several albums, they decided to stop performing. It had now been about 11 years since their debut and they had almost no days to rest. The band performed on TV both live and recorded, toured the nation for live concerts, and stayed in the studio to record new songs. During this time, Yoko had almost no privacy and needed to rest for a while until she regained her energy to perform again. Each member had different reasons to stop performing; for Yoko, it was to improve her mental health. Her psychiatrist diagnosed her with Autism Spectrum Disorder, Attention-Deficit/Hyperactivity Disorder, and Obsessive-Compulsive Disorder. None of her symptoms were critical; however, the psychiatrist recommended she rest for a while.

Yoko had noticed a power struggle among band members. Two other members and management staff usually decided the contents of their work. The band leader had been the primary lyricist, composing all of their 250 songs, except the two songs Yoko wrote and composed. Furthermore, although two of the guitarists stood behind her on stage while she performed, she felt they had more control over their stage performances, especially the bandleader. It had been more common for him to cut her talk on the stage and take control of the conversation. She felt upset but needed to continue to perform well after those incidents.

The band members, supporting members, management staff, producers, and all of the entertainment professionals were male except a personal assistant. They often commented on Yoko's looks, which had made her very self-conscious. As a vocalist of a popular band, she thinks it is important to not look fat and unhealthy. At the TV shooting, one of the male artists commented on her, saying "Long time no see! You look fat now!" She reacted to the comment and felt she may need to go on a diet. She does not have unhealthy eating behavior but she constantly thinks about her weight. She also goes to a salon once a week to receive facial and body treatments to avoid looking "plump." She has not done any cosmetic surgery and has no intention to proceed with surgery in the future. Yoko just wants to make sure she does not look fat. In the past, Yoko tried hard to

DOI: 10.4324/9781003089629-15

look like a happy, cheerful girl. However, she is not in her 20s anymore, and fans may expect something more from her, but she didn't know how to find or create a vision as an artist.

After almost two years, her condition had improved and she felt ready to perform again with the members. Even though the environment surrounding Yoko had not changed at all, including the entertainment culture, her relationship with band members and management staff, or her lifestyle, her motivation was the same: earning money. Yoko lost some money during those two years and started singing again to earn money. However, she had already earned enough money to live without further earning. As the band restarted, it was a chance for Yoko to organize her thoughts toward a career as a vocalist.

Even though her condition improved, the psychiatrist recommended that Yoko work with a performance psychology consultant who offered a cognitive-behavioral approach. Yoko asked her management staff to find a consultant, and is excited to work on her psychological skills to enhance her performance. She feels she needs to change many things to become a better signer after resting for two years. Her major concerns are about emotional control, steroid use before a live performance on TV, and feeling pressure to please others including band members, staff members, and fans.

As Yoko goes back to a popular artist's life again, she knows she needs to change her life. She cannot wake up late in the morning or continue to sleep in. She leaves home in the afternoon for work and comes home close to midnight or even later. She usually stays in the studio, spending time in meetings, vocal lessons, recordings, and photo shoots, among other things. Sometimes, she goes out for a drink or to party with the staff after work, not because she genuinely wants to, but more to please her staff members. She finds it difficult to sleep because her arousal level is too high. Instead, she spends her time training and writing lyrics (to remind herself) or singing songs at home. Also, she cannot stop listening to podcasts (about spiritual life) to make her feel better, but they also make her worry about her life. She is also unable to venture outside for shopping or meals due to her popularity; she often needs to prepare meals or her staff needs to provide her meals. She also wants to exercise to be physically healthy and sing well on the stage. However, she does not have enough time to invest in eating healthy and exercising regularly.

Yoko works with two managers very closely. The male manager communicates with the management company and decides the schedule for Yoko. He oversees Yoko's condition and controls her workload. The female manager is the one who observes Yoko's physical and psychological condition closely. She sometimes notices Yoko is excited about her projects and carries full energy all day. She becomes talkative, sings too many songs, tries to communicate with many staff, walks around in the setting, and eats snacks all day. Yoko does not take the time to rest because she does not want to disappoint the people around her. It usually takes nine to ten hours to complete the day regardless of the project, such as recording songs, photo shoots, shooting a TV show, or singing live. By the end of the day, Yoko is exhausted.

A typical schedule for a live TV performance consists of the manager picking Yoko up by 1:00 PM to go to the TV studio and taking her home by 11:00 PM. Yoko rehearses at 5:00 PM and the live performance is at 9:00 PM, usually lasting between four and eight minutes. During her stay in the dressing room, she needs to change her clothes and put make-up on, but she has no consistent routine for this time. This means Yoko spends time on things not necessarily important or relevant to her optimal performance. She chats with random people, eats snacks, sings, and writes lyrics repeatedly.

Yoko tends to prepare too much in terms of singing and remembering lyrics. While it depends on the schedule, she usually performs two to three times a week. She also goes to voice training once a week. During studio recording, she sings five to six songs, which generally takes a couple of hours. She worries about the recording, so she starts listening to the music on her smartphone and singing repeatedly for a couple of days before recording. When the time comes for the recording, she does not have her voice left. If her performance is a TV recording or live on TV, she worries about her performance and sings the song(s) many times at home and in the dressing room. Also, even though she can follow a prompter on the stage, she is afraid of not remembering the lyrics. Thus, she repeatedly writes the lyrics the night before at home and in the dressing room as she prepares for the stage. She feels extra pressure for live TV because she is the only one performing while all the source sound for the instruments is prerecorded. As she gets ready, she often becomes anxious about not performing well and tends to take a steroid pill (anti-inflammatory) which makes her feel better about her throat. However, she wants to be able to sing and perform without taking the steroid pill in order to continue a healthy career.

Thought Questions

Case Conceptualization

- What information do you need prior to beginning work with Yoko in order to enhance your effectiveness?
- What key aspects of Yoko's past are key to understanding her current circumstances?
- What are the important elements for her as she develops a meaningful career?

Ethical Considerations

- Do the facts in this case suggest any "red flags?" That is, do you notice the need to consider clinical referral, limits of your consulting competency, competing stakeholders, etc.?

- What could be potential signs in Yoko's behavior which may indicate consideration for cessation of performance consulting and sending her back to the care of her psychiatrist?
- How do you approach Yoko's behavior of taking an anti-inflammatory pill occasionally before performing, along with medication for her disorders?

Diversity Awareness/Reflection

- What are your initial thoughts regarding the culture involved in this case? These may be related to the performance arena or the individual performer.
- What can you do to be culturally competent in this entertainment setting?
- How might Yoko being a female member in the band impact her? (She is a vocalist and band's popularity depend on how she sings, looks, and talks.)
- How might you work with her management staff to maximize her performance and minimize her fatigue?

Intervention Planning and Evaluation

- Identify the potential issues that may be affecting Yoko (these may be personal and/or performance related). From your list, prioritize the top three issues that you believe are having the most immediate impact.
- What are the criteria to evaluate the success of your work with Yoko?
- Does Yoko's age (35 years old) impact your intervention design? If so, how?
- What are your ideas to help Yoko to continue having a successful music career?

13 Acting

"One Actor's Journey: Moving Beyond a Failed Audition"

Jacob Jensen

UNITED STATES

Kelli has been acting from the time that she was a young girl participating in community theater, local commercials, and school productions. Her parents had immigrated from Taiwan before she was born and settled in a midsized Midwestern city in the United States. Her mother taught Mandarin at a private school, and her father was an engineer. Although her hometown is not a huge market in terms of acting, there were a number of opportunities for her to pursue her acting career while living at home and attending a performing arts high school close to her home. While in high school, she often attended acting schools during the summer in New York (NY), London, and Los Angeles (LA), which allowed her to be a part of summer play productions, as well as work and network with acting coaches and actors from around the world. Although much of her acting experience growing up was in theater productions, her dream was to be a television and film actor and move to LA full time. Kelli was well known in her hometown as a talented actor who booked leading roles in many theaters and school productions. She booked many local commercials and even a few national commercials and was a "big fish in a small pond" in terms of her recognition as an actor. In addition to being told that she had a lot of talent for acting, she has always worked very hard on her skills and has had no doubt that acting was the career she wanted to pursue. Her parents had always encouraged her pursuit of acting, especially when she was young, although they also emphasized the importance of doing well in school and pursuing other interests. In addition to acting, Kelli was also involved in ballet beginning at age six, but during high school, she gave up performing in dance and only continued it to stay in shape.

During her senior year of high school, Kelli faced the decision to move to LA or NY after high school or pursue a theater or film acting degree in college. Some of her high school friends and others she had met in the summers moved to LA and NY to pursue acting full time. She also had some fairly high-profile agents and managers interested in representing her in these markets if she were willing to move. However, Kelli's parents were against this move and wanted her to get a college degree first and then make a decision about moving to LA full time after that. She decided that it would be best to follow her parents' advice and enrolled in

DOI: 10.4324/9781003089629-16

a university in the Midwest that had a good reputation as a top collegiate theater program. Kelli attended the university, majoring in theater. Even though she did well in her classes and university theater productions, she felt she was not pursuing her dream of becoming a tv/film actor. She saw some of her friends starting to book some roles in films and tv shows and felt that she was falling behind. She also felt that the acting classes she took in college were somewhat limited in only teaching one acting method, whereas before college, she had studied many different methods and drew from a variety of approaches and acting techniques and philosophies. When she approached her parents about this, they continued to encourage her to finish her degree, which she did in four years.

After graduating from college, Kelli was still determined to move to LA to pursue her acting goals, but felt burned out, frustrated, and lacked confidence in her acting skills. She also felt that she had spent so much time in theater that she was behind when acting on camera for tv/film auditions and roles. After moving to LA the summer after her college graduation, she also found that agencies were no longer interested in her. She had not done any television or even commercial acting in college, and her credits for the previous four years were mainly connected to the theater. Several agents told her that she should go to NY because her theater background would serve her better there. Without representation in LA, she found it very hard to audition other than for small student productions or non-union commercials. In order to pay rent and her bills, she had to get a job teaching ballet at a dance studio. Her parents were pressuring her to find a "real" job because they did not think she could support herself as an actor in LA.

For a few years, Kelli continued to struggle to find an agency to represent her and to support herself in LA. After three years, she got a small boutique agency that represented her across the board (commercials and film/tv roles). She booked a couple of SAG national commercials that allowed her to quit teaching ballet and support herself as an actor. This also allowed her to join the main actors' union, SAG/AFTRA (union productions), which gave her more credibility as an actor and greater financial stability. She then was able to sign with a larger theatrical agent (tv/film), which provided her some bigger auditions for series regular roles and network pilots. She started to book a few guest-star credits, had agency representation, and was supporting herself as an actor. However, she still had not booked a significant role in a film or as a series regular on a tv show, and her parents still referred to acting as her "hobby," which bothered her. She felt that her parents and family members would not take her acting career seriously until she booked a major role in a film or television series.

One day her agent called with news about an audition for a role in a 30-minute single camera comedy TV pilot that was being produced and was already being considered by a major network. The pilot focused on a Chinese family who emigrated to San Francisco, and her agent said it was a "perfect" role for her. It was a series regular role, and she would be required to have a Chinese accent. The casting director had seen a guest star role that Kelli had on another show and was pushing her for this role. Kelli worked with her acting coach on the audition and worked hard on her Mandarin accent (Kelli was born in the United States and speaks with no accent, although she does speak conversational Mandarin and can do an accent). She even called up her mother to listen to her and imitate and practice her mother's accent, which is close to what the casting director told her they wanted for the role. Kelli made it through the first two rounds of auditions and was called back for a chemistry read with one of the lead actors (who had already been cast for the pilot) in front of the casting director, the writers, and producers of the show.

When Kelli arrived at the chemistry read, she signed in and then went to the waiting room. Her agent had told her that the role came down to her and one other actor. As she was waiting for her chemistry read, she saw another Asian American actor walk out of the audition room, and Kelli immediately recognized her as one of the most well-known Asian American actors in Hollywood with many films and tv credits to her name and who had been nominated for an Academy Award the year before. Kelli realized that this role had come down to her and this other actor who was far more accomplished in terms of credits and was much more of a household name. When Kelli entered the audition room a few minutes later, she was overwhelmed with anxiety and nerves as she saw at least 12 people in the room and kept thinking to herself how this role could change her career and life forever. One of the producers then told her that they had decided that her character was from Taiwan and that they wanted her to do a Cantonese accent. Kelli told the producer that they speak Mandarin in Taiwan, not Cantonese, but he just laughed and said, "It doesn't matter; Americans won't know the difference." As Kelli started the audition/chemistry read, she tried to do a Cantonese accent instead of the Mandarin one that she had practiced, and what came out was "such a mess" of an accent that the other actor could not even understand her. The casting director stopped her after a few lines and told her just to go back to the accent she had done in the previous rounds of auditions, and Kelli was mortified that she could not make the adjustment and stammered her way through the rest of the audition and completely lost the voice and accent she had used to get to the chemistry read. She was also furious that she was being asked to do a Cantonese accent when her character was supposed to be from Taiwan and speak Mandarin.

Kelli left the audition room, realizing that she had bombed the audition and her chances of booking the role. Her agent called her the next day and asked what had happened. They had offered the other actor the role, and her agent wanted to know why she had "choked" so badly in the room. The television series was going to be filmed before a live audience, and the producers were worried that if she had frozen under the pressure of the audition room, what would happen in front of a live audience. Kelli's agent told her that she could not continue submitting her for bigger roles if she did not trust that Kelli could deliver under pressure. When Kelli called her parents and told them what had happened, they told her that maybe it was a sign that she should quit acting and find a "real" job.

Kelli's agent had met me, a sport and performance psychologist, at a workshop that I had given to agents and managers on mental training and mindfulness work for actors. The agent gave Kelli my name and number and said that she needed to book a few sessions with me or else she would be dropped from the agency. When Kelli called me and explained what had happened at the audition, she was almost in tears talking about how she "had bombed" her big chance. She feared that she might never get another chance and that her agent might drop her, and she worried because her parents thought she should just quit.

Thought Questions

Case Conceptualization

- What information about Kelli's history as an actor could influence your understanding as a sport/performance consultant of her current situation of "bombing" a major audition?
- How do factors such as family pressure, relationship with the agent, and others influence how you might approach this case?
- What factors and circumstances made the "stakes" particularly high for Kelli in this audition? Why do you think that Kelli reacted so negatively to being asked to do a different accent than the one that she prepared?

Ethical Considerations

- Kelli was told by her agent that she would be dropped if she did not seek help from a sport/performance psychologist. How would you respond to or communicate with the agent if she asked for an update on how Kelli is doing?
- What other individuals could you potentially talk to in order to gain insight into Kelli's acting career (e.g., acting coach)?
- What could you do to gain a better understanding of how Kelli acts or performs under pressure (acting class, theater, self-tape audition)?

Diversity Awareness/Reflection

- How familiar are you with acting culture, the power dynamics that exist between agent and client, as well as the casting and audition process?

- What additional background information would you like to know about Kelli?
- What sociocultural factors could have influenced Kelly's reaction in the audition room?

Intervention Planning and Evaluation

- What about Kelli's background as an actor and her journey to having a big audition for a tv series in LA impact some of the beliefs about herself that she may have taken into the audition room?
- How would you help Kelli bounce back or recover from being "devastated" from this audition? How would you help Kelli evaluate progress as these types of auditions do not come around very often, and they will not come again until her agent is confident in submitting her for these types of roles again?
- What types of interventions or training do you think would be best for assisting Kelli to continue to progress as an actor and to reach her goals?

14 Culinary

"The Proof Is in the Pudding!"

Frank Eirik Abrahamsen

NORWAY

Sandra is a 21-year-old chef working at a Michelin star restaurant in Norway. In addition, she has been a member of the chef's senior national team for nearly two years. She began her training as a chef at the age of 15, and it was evident to all around her that she possessed both talent and ambition. As a young upcoming chef, she was also a member of her country's junior national team for a total of two years prior to being promoted to the senior national team of chefs. The restaurant where she is currently employed is situated in a remote area. Meaning, to attend training sessions with the national team, her commute is a full day's work – each way.

Her area of expertise as a member of the national team is desserts, in culinary language her title is "pastry chef." When it comes to her paid job, the majority of her work in the restaurant focuses on main courses. Tim, who is one of her colleagues at the restaurant, is a former contender of the national team. Due to a lack of progress in his creations while on the national team, combined with some difficulties in his cooperation with the national team coach and in fulfilling his obligations, he was dismissed from the team. Sandra was originally a "commis" (i.e., a junior chef) with the senior team, and was promoted to Tim's position following his dismissal. Currently, Sandra is working as a chef making main courses in the Michelin star restaurant, while simultaneously performing as pastry chef with the national team, in Tim's former position.

Preparing food at the Michelin star level as well as the national team level has proven to be quite demanding. The food must be prepared in a perfect manner, the ingredients must be of the highest quality, and the presentation of the food must be flawless. Needless to say, the taste of the food must also be outstanding. It is not only necessary for the food to be in perfect order to win the Olympic medals, but the chefs must also demonstrate a high level of technical skills while preparing the food. In demonstrating such skills, one is expected to create complex creations while making the effort seem simple. This complexity applies to several Michelin Guide restaurants, as customers desire complex tastes with several layers of depth, yet both complexity in her creations and simplicity in the effort put forth apply when it comes to Sandra's skills with the national team. At the same time, Sandra needs to master two varied types of food –

DOI: 10.4324/9781003089629-17

preparing main courses in the restaurant and preparing petit fours and desserts for the national team.

The Culinary Olympics are hosted every four years (typically in February). This time of year is one of the most hectic periods for chefs in Norway. Christmas parties are a long-standing and important tradition for Norwegians and are therefore of great importance in terms of revenue for the restaurants. Sandra, like most national team chefs, conducts her national team training during her spare time. This requires goodwill from the restaurant owners and good working relationships with her colleagues. In order to conduct all the training necessary for winning the Olympic medals, a chef must undergo a vast amount of training, both individually and as a team. Most restaurant owners consider it an honor to "provide" chefs for the national team. However, this has proven to be challenging during November and December, as the restaurants require their entire staff to meet the demands of this hectic period of the year. Naturally, every four years, this coincides with the very hectic period for the national teams in preparation for the Olympics. The events below took place in the four months leading up to the Olympics in early November and are narrated here as they were happening.

Too much is going on and Sandra feels that she is faced with several challenges. Her training as a pastry chef requires more spare time, which puts a strain on her relationship with both her employer and colleagues at the restaurant. In addition, she is already exhausted from the hard work at the restaurant and has little-to-no energy left to develop her skills at the level the Olympics require in the months prior to the competition. An additional challenge is that her rapport with Tim has suffered a full breakdown. He feels that Sandra backstabbed him by accepting the promotion with the national team. The fact that he was replaced by a junior chef with a different expertise than his, while he has been working at a Michelin star restaurant, of course, had a negative effect on his reputation. Recently, he complains often about Sandra receiving, what he perceives as, special treatment from the restaurant owner in the period leading up to Christmas, when according to him she should manage to fulfil her obligations at both the restaurant and the national team like he did.

In short, the situation is affecting Sandra's motivation and stress level. Her motivation is starting to diminish. She is feeling trapped and has extraordinarily little autonomy in her day-to-day activities. A typical day begins around 6 am when she goes to the restaurant to make the necessary preparations for the meals according to her responsibilities. The restaurant`s dinner service concludes at 10:00 pm, which is when she cleans her station. Sandra is usually back at her apartment around 11:30. This is her normal regime most days, except for Mondays when the restaurant is closed. The national team usually trains together once a month from Friday morning to Sunday afternoon, while its members are expected to train on their assigned creations in between these training camps. As the Olympics are approaching, the team has scheduled training camps for every third week, from Friday to Monday afternoon. Typical travel for training camp, means Sandra must first drive for several hours to the nearest airport, take a small plane to a larger city, before the final flight takes her to the training camp, and similarly on her way back. Because of plane connections, she is sometimes absent from the restaurant from Wednesday afternoon until Monday afternoon. Sandra feels as if the work has taken over her entire life and that she has become a back-seat passenger in her own life. The strained work relationship with Tim in the restaurant has also taken its toll, making her feel that there are very few places for her to rest, physically and mentally. On top of everything else, she is finding less of a sense of accomplishment and expertise with dessert preparations on the national team recently, because perfection seems to be the only accepted standard – anything less is regarded as a failure. She is expected to perform perfectly – while getting criticism for sub-par standards and little or no feedback for good work.

There are two topics that make Sandra feel especially worried and stressed. First, the pressure to perform is growing as she was awarded the role her colleague, Tim, had specialized in and she had not – yet, at least. She feels the pressure to outperform Tim to defend her position and promotion within the team. The fact that a specialized chef was dismissed from the team, worries her – is she next in line? She feels she knows where she is heading as a highly skilled pastry chef; however, she is certain she will not make it there in time for the Olympics. Even when the other chefs on the national team and the head coach are satisfied with her work, she keeps making minor adjustments to the recipes in a quest to improve them. One of the problems related to this is that she never fully trusts her own skills and creations. An additional problem is that she never reaches the level where her abilities are fully automatized. A lack of automatization, might lead to second-guessing the delivery during the competition – something she will not have time to deal with. The catch phrase "paralysis by analysis" worries the national team coach before the Olympics, and he is particularly concerned about Sandra.

Following New Year's Eve, Sandra calls in sick to the restaurant and the national team. The coach of the national team calls her back. In the beginning of the conversation, she sounds irritated about not being left alone. However, eventually she understands that the coach is worried about her and wants the best for her. She conveys to the coach how exhausted she is and that she is struggling to focus on work and is having difficulty sleeping. During their conversation Sandra talks openly about all the issues she's facing, and the coach praises her for the openness. Sandra and the coach have a long conversation and agree to continue the conversation in the following days in order for the coach to help her solve these issues.

In agreement with Sandra the coach calls you, the mental coach of the national team to discuss this case. Both you and the coach are worried about her well-being and her ability to meet all the demands during the two months leading up to the Olympics, as well as her performance during the Olympics. The coach and Sandra discussed whether it would be better for her to leave the team, but she conveyed that she has a strong desire to stay on the team and that she would feel she is betraying her teammates if she were to leave at this point. This is not an option for Sandra. The coach feels the same way as Sandra; she should continue with the national team as her technical skills are exceptional. In addition, she possesses certain personal qualities as a team member which will be challenging to replace on such short notice. Her professional and personal qualities were among the reasons why the coach chose her as the pastry chef. The coach explains in an honest manner how there were no other options for him and that she was, and still is, his first choice regardless.

You know Sandra to a certain extent from working with mental training in group sessions and team preparations, such as goal setting and team dynamics. Now, there are only seven weeks left before the Culinary Olympics, with three national team training camps ahead, the first one just ten days from now. The coach and Sandra ask for your assistance.

Thought Questions

Case Conceptualization

- What do you consider to be Sandra's major challenges at the moment?
- Which factors do you think are controllable for her both short- (in next two months) and long-term?
- Jobwise, what may be alternative solutions to Sandra's case and how would these affect her potential career options?

Ethical Considerations

- You are employed by the federation which has been involved in national team members' selections/deselections, what boundary- and consultation-related issues should you beware of?
- Are there any topics that you should be especially concerned with in Sandra's case?

Diversity Awareness/Reflection

- What contextual knowledge of the haute cuisine culture do you have to assist you and your work in this setting? What missing contextual pieces would you like to obtain more knowledge regarding?
- How may the fact that Sandra has not conducted any formal training as a pastry chef affect her association with her older, trained colleagues, if she succeeds at the Olympics?

Intervention Planning and Evaluation

- What do you consider to be the sources of stress (stressors) from Sandra's environment, and what stress sources are self-induced?
- What determinants would you use as evaluation criteria to evaluate the success of your work with Sandra?

Section 4

Trends in Consulting

15 Corporate Consulting

"Seeking Next Level Leadership"

Adam H. Naylor

United States

Grace is an emerging corporate leader who has chosen to take advantage of a mental performance benefit offered by her company. As someone that sits at the executive-level, she leads a corporate division responsible for multi-million-dollar businesses and is responsible for a large workforce and their leadership. Initially, this consulting arrangement was arranged to be all remote consultations, yet due to the company's enthusiasm for such resources for its premier leaders, travel and time for in-person meetings would be supported at the client and consultant's discretion. Upon the first call with the consultant, Grace's enthusiasm toward wrestling with complex business challenges and her commitment to build and serve high-performing teams is evident. She is quick to say, "I am always trying to learn how to up my leadership game and am always reading, filling leadership books with sticky notes and highlighting."

Beyond being a corporate leader, Grace is married and mother to two children. Sally, her daughter, is nine years old and enjoys school, swimming, and golf. Peter, her son, is five years old and has a love for baseball and all things superhero. Grace is 50 and finds the comfort to share in the first meeting that she and her husband waited to have kids out of concerns that motherhood would derail her path up the corporate ladder. Her husband Sam, is a corporate lawyer that logs long days and hours in the office. They are a committed and caring team, yet Grace does lead the roles that fall traditionally on mothers – meal planning, back to school shopping, kids' activities, schoolwork planning, and the rest. Homelife is a carefully choreographed dance of schedules. To assure adult work can get done, kids can embrace recreation, and the family can steal quality moments together a few times a week whether at the ball field or the dinner table. This is all supported by the part-time nanny that, in discussions, sounds much more like a family member than "the help." It sounds like it is a family full of laughter, love, and regular chaos.

Grace's personal narrative is filled with drive, insightfulness, and empathy. While she has been part of a forward-thinking company for over two decades, it would be remiss to think that she has not faced gender disparities and invisible obstacles. When she entered the workforce, white men in suits were at the front of the line. She navigated this landscape

DOI: 10.4324/9781003089629-19

with grit, sacrifice, and poise. Having arrived at the elite leadership level, she finds herself less concerned about being accepted as a smart, strong woman. Chauvinistic foolishness rears its head at times but having navigated the gauntlet to get to this level with her values and voice intact, Grace feels able to shape culture and strongly promote an equitable and inclusive workplace. Now in a leadership role, it is the millennial "bro culture" that feels like the bigger challenge to inclusive teams early in this century that she wants to tackle. She is proud of her efforts to achieve such professional heights and hopes to be a positive role model for the women that follow her. She makes it clear that bringing diverse voices to the table is core to her beliefs about how high performance is achieved.

In the midst of work and family busyness, Grace strives to check all of the wellness boxes. While almost always putting in sixty to eighty hours of work a week, she finds herself sleeping seven hours a night on a regular basis. Despite being on planes and in a hotel a few days every other week, her diet is heavy in chicken and fish, and, unlike her colleagues on the road, she limits her alcohol consumption, having only a glass or two of wine with dinner if at all. As far as getting herself physically moving, she is a fan of her virtual cycling classes. She is fairly committed to getting a ride in regularly, as long as the bike in her home gym is not multi-tasking as a laundry rack or roof support for a kid's fort. Grace seems to be doing a pretty good job taking care of herself in the midst of the hustle and bustle, but she does say that it is an aspect of her life that she would not mind improving a bit.

The intake meeting with Grace provided all of this information, and the meeting was concluded by setting some goals for the mental performance consulting work. Three initial goals were determined for the consultation: (1) find a bit more life balance, (2) find and use her leadership voice with peers and more senior leaders more often and more comfortably, (3) take advantage of the mental performance consultant's science-to-practice knowledge of leadership to simply grow as a leader. As often happens at the beginning of a consulting relationship, the goals and objectives evolve and get refined upon reflection and deepening of the working alliance. This was the case in this relationship where the life balance and general leadership learning goals were determined to be "nice to have," but not necessities. Instead, learning how to cope with a seemingly divisive peer and having a sounding board for day-to-day leadership challenges arose as more valuable uses of performance consulting time.

At the start of the second meeting, Grace shared the story of a peer that had been difficult and seemed deceitful toward her over the past two years. This all occurred during her process of being promoted to higher leadership roles and involved stressful discussions with human resources and company leadership. It seemed to reason that there were some jealousy and power dynamics involved in the behaviors directed toward Grace. To exacerbate the matter, the peer is also a striving female leader. This has led to some cognitive dissonance as Grace is at the forefront of championing the promotion of women in the workplace, yet has no patience for poor teamwork and leadership regardless of gender. At the beginning of this consulting relationship, this adversity had faded to the background of day-to-day work. Yet, emotional bruises had happened, and seeds of distrust had been sown.

Every week, Grace is responsible for long-term strategic planning and budgeting, shaping teams, mentoring the leaders of the various teams, and serving as one of the public faces and voices of the company. This requires navigating a wide variety of personalities and the ability to find the right tone and tenor for all communications. It also asks for skill in sharing a vision and expertise in a manner that media can embrace, and the global marketplace can get behind. The time commitment for all of these professional roles could be unending. Grace wonders if she could delegate better. Grace wonders if a 60-plus-hour work week is foolish. Perhaps it is time to find that often talked about and ever-elusive "life balance."

As impressive and driven as Grace is, she is not pleased with her use of her leadership voice. However, this is a nuanced concern of hers. She has tremendous followership, and over the entire course of her career, she has always brought novel perspectives forward during projects. Furthermore, she would have never risen to the leadership level she is at without the respect of peers and more senior leaders. Her perspectives and ideas are not only valued within her organization but also seen as essential to the business's success. On the surface, it seems absurd that Grace questions her leadership voice. Her self-reflection leads her, however, to appreciate that the energy she presents when paired with her understanding of both technical and marketplace challenges, can leave some people on their heels. Wide-eyed younger employees are excited by the energy and jump on board. Peers and elders could feel threatened or even confused by the energy. These are perhaps more internal, personal concerns than reality. Yet her respect for more veteran leaders in the company is very real, their intelligence, and what they have accomplished. This most specifically leaves her silent sometimes when she would like to speak up.

Upon getting to know Grace, it is easy to be in awe of her accomplishments and how she navigates the world in which she lives. She is a loving mother, forward-thinking leader, and firmly committed to succeeding in ways that lift up both her company's business and her company's people. As is true with exceptionally high performers in sport, the arts, and business, obvious "weaknesses" in their approach to performance are very limited. Grace has achieved an elite professional status that few of her peers will find. Yet, she has a thirst to continue to grow and learn. Throughout conversations and examination of her goals, it becomes evident that dialing in her performance perspectives and mental skills will likely reap benefits to her in both her professional and personal life. She is excited to have a sounding board and thought partner on the mental game.

Thought Questions

Case Conceptualization

- Is there anything that you need to know about the industry in which Grace works in order to enhance your effectiveness as a consultant?
- Is there anything specific about the corporation in which she works and the corporate culture that would benefit the consultant for playing their role well?

Ethical Considerations

- Are there elements of this case that would bring into question a consultant working in or out of their competency? How would it be determined if this is an appropriate consulting fit?
- Should there be any concerns regarding the company providing this benefit and its role in the consulting relationship? If so, how could these be navigated?

Diversity Awareness/Reflection

- What are your initial thoughts regarding cultural considerations in this case?
- Will the gender of the practitioner impact ability to provide effective consultation in this case?

Intervention Planning and Evaluation

- There are many potential goals for this consulting relationship. What topics do you see taking the highest priority in the consulting relationship?
- How will you know if the goals of the consulting relationship have been achieved?

Additional Thought Questions

- What is your conceptualization of work-life balance? In your opinion, is this something that deserves significant focus?
- There seem to be many gender issues at play in this case. Grace has not explicitly stated a desire to address any of them directly. How front of mind and front of conversation ought the consultant keep these matters?
- Workplace grievances and conflicts are stressful and emotional. Although the matter with her peer has been "resolved" by company standards, is it appropriate for the consultant to consider the emotional nature of the experience, something that affects future performance?
- How do you coach someone to finding a greater "leadership voice?"
- Does Grace need a mental performance coach?

16 e-Sport

"Ninjas in Pajamas"

Doug Gardner

United States

Dota 2 is a Multiplayer Online Battle Arena (MOBA) game, in which individual players control a single character as a part of a team competing against another team of players. In simplistic terms, Dota 2 is a very complex version of a five-on-five "capture the flag" game. Winning involves destroying the base of the opposing team by fighting each other through a map that consists of top, middle, and bottom lanes and two jungle areas. Players amass gold to purchase different actions, potions, and other concoctions which allow characters to grow in strength as the game progresses, build team strategies to take in-game objectives, and establish leads to win.

Dota 2 has over 120 different characters, called heroes, to choose from in the game. Each hero has unique attributes, strengths, and weaknesses when paired against other heroes from the pool. In competition and practice, teams conduct drafts against each other, where they ban and pick heroes in a back and forth manner, banning certain champions out of the selection process while leaving others open to choose or be chosen by your opponent.

One of the most challenging aspects of the game is that the game developer will make in-game changes every few weeks called patches and/or meta changes, requiring all teams to adapt, adjust, learn, and apply strategies based on these changes. All of a sudden, certain heroes "play" completely differently than they did last week, which then impacts match-ups against other heroes. At times, this is a benefit, but at other times it causes great players to become weak. Specifically, changes in the game impact the ability to adapt a player's play style and ability to play multiple heroes at a high level. This, in turn, can have negative effects on both individual and team performance.

The structure of competition at the professional level is similar to professional golf and tennis; winning regional tournaments allows teams to qualify for major tournaments played all around the world in front of thousands of fans both in-person and online. Tournament hosts fly teams to venues to compete and the only way to make money as a professional is to win or finish in the top tiers of tournaments. Prize money is divided among players. Each year culminates with a grand final event, called "The

DOI: 10.4324/9781003089629-20

International" and the team that wins can walk away with as much as $30M dollars.

Social norms in esports are quite different than in traditional sports and the Dota 2 ecosystem is a colorful environment filled with people called "Dotaheads," which was adapted from Grateful Dead fans, Deadheads, who show tremendous devotion to Dota 2. Unlike some game titles in which there are leagues and franchises owned by large investor groups, many Dota 2 teams are their own entities; free to choose players, staff, coaches, and locations at which to live and train.

These elements lead to frequent player movement from team to team, often occurring between major tournaments throughout the year. Several teams are player led and do not have head coaches. In addition, players come from all around the world and teams are often comprised of individuals who have socio-cultural differences or potential language barriers, both of which can impact individual and team performance.

The Dota 2 season lasts 11 months and there are short periods of time, usually after major tournaments, during which teams take time off and get away from the game. Teams will come back together and "bootcamp" over several weeks, where they intensely prepare for and play in qualifying tournaments which, hopefully, culminate in major events and money-earning opportunities. In bootcamps, players and teams usually travel to a location close to other teams, often spend anywhere from six to ten hours a day scrimmaging against teams, and practice playing their heroes individually with and against random players online.

Two months into your tenure as a performance consultant with an esports organization, the Dota 2 team returns to bootcamp from a short break after failing to qualify for the year's first major. The team was originally made up of five friends, all from the same country, but two decided to leave the team due to philosophical differences in play style. The team added two younger and less experienced players, one from North America and one from South East Asia, and a strategic coach to help them with strategy, preparation, and to resolve conflicts.

In addition, this team is sponsored by an esports organization who pays them a salary to represent their brand. The organization recently opened one of the industry's first training facilities, which means the players now have to train and practice away from the team's house. Players now have obligations to stream on various social media platforms and to commit time to creating content to interact and grow the organization's fan base.

The team's first day on campus was a memorable one to say the least. They were supposed to arrive for an 11:00 AM meeting with you and the CEO, eat lunch, and begin training for the next series of qualifier tournaments. Around 2:00 PM, the team strolls onto campus looking disheveled; one player is literally wearing pajamas.

Up to this point, you have little interaction with the team as they were competing in Europe for a spot in the first major before you start to work with them. During the team meeting, the players were aloof, indifferent, and did not want to hear from you, the CEO, or anyone else. They just wanted to practice, finish up, and go back home. Regardless, you outlined your role in the organization and explained to them how you hope to earn their trust and respect with the goal of helping them regain their status and success on the international stage.

You felt their apprehension, as the core three players, who were a tight group, now had to manage integrating two lesser experienced players, work with a strategic coach they did not want to have, and determine how to navigate a new corporate work environment with external expectations to meet beyond that of competition and winning. They were now a part of the corporate machine of esports and had responsibilities, such as streaming hours, content shoots, and promotional videos to make for the organization.

Practice consists of scrimmaging other Dota 2 teams online anywhere between four to six games a day, with each game lasting approximately 40 minutes. After each game, the team replays the game, doing what are called "vod" reviews that last about 30 minutes, so they can learn from mistakes, establish game play styles, and glean information with the goal of team synergy and improvement. This process usually lasts between five to seven hours a day, depending on how many scrimmages they schedule.

When team practice ends, the players go back to their team house, eat something, and then continue playing the game, but in a more unstructured format. Often, individual team members stay up late playing with random players online, forming random teams and mindlessly playing until they decide to go to sleep sometime between 3:00 and 4:00 AM. Such is a day in the life of a professional Dota 2 player.

The team does not have any of the habits or established routines that are the norm in traditional sports. With no formal coaching or external discipline, it is up to the players individually and collectively to set up their day-to-day work. Nobody makes them practice, nobody makes them go to sleep, nobody makes them get up, eat something healthy, exercise, or prepare as if they are professional athletes. The core of the team has been together for over three years, making The International two years earlier, finishing seventh, and earning a good amount of money. Now, they had to rebuild in the middle of the season and the lack of leadership, team focus, and direction was clearly hampering the team's ability to be successful.

You realize a few things as you continue to observe the team over time. When someone tries to impose structure on them, they rebel, as nobody from a Dota 2 team wants to be told what to do, let alone hear how they should prepare, especially from a "non-endemic." Team leadership had to come from within, yet nobody was stepping up. From a cultural perspective, the concept of a singular leader was also foreign to these players. Finally, with two new players from two different cultures, along with an inexperienced coach, the team was in big trouble.

Clearly, the team was a mess and you knew that an intervention was necessary. You also knew that you could not come in and start making demands of the players and team because you just started working for the organization and had no authority over the team. You watched them bootcamp for four weeks and everyday was like Groundhog Day, with the only consistency being inconsistency. You did not need to be an expert in Dota 2 to see that the team was struggling on many levels, and sometimes you have to let teams and/or individuals lose before they become open and receptive to feedback and teaching.

Ultimately, the team failed to qualify for the second major of the year and now had gone four months without winning or making any money. They were feeling pressure from the organization that was counting on this team to qualify for the end of the year championship, The International, and under the threat that the CEO would disband the team all together if they did not start showing some semblance of success over the next few months.

Thought Questions

Case Conceptualization

- What information do you need prior to beginning to work with this individual/group in order to enhance your effectiveness?

Ethical Considerations

- Do the facts in this case suggest any "red flags?" That is, do you notice the need to consider clinical referral, limits of your consulting competency, competing stakeholders, etc.?

Diversity Awareness and Reflection

- What are your initial thoughts regarding the culture(s) involved in this case? These may be related to the performance arena or the individual performers.

Intervention Planning and Effectiveness

- Identify potential issues that may be affecting the performer(s) (these may be personal and/or performance-related). From your list, prioritize the top three issues that you believe are having the most immediate impact.

Other Questions for Consideration

- Where do you start, what do you do first?
- What information do you have and what information do you need to help the team perform at a high level in a short period of time?
- How do you get buy-in from the team?
- What structure would you create to maximize daily preparation and performance?

17 Action Sport

"The Apprehensive Action Sport Athlete"

Sarah Clifford

UNITED STATES

Sonya is a young adult professional action sport athlete originally from a small town in the northeast, and an only child to her single-parent mother. From a young age she was extremely active, wanting to try everything. She played soccer, swam, took riding lessons at a local stable, played t-ball and basketball, wakeboarded at a nearby lake in the summer, and snowboarded many weekends in winter. She particularly enjoyed being on wheels, whether it was on a bike, skateboard, or rollerblades. She loved getting out of her comfort zone and was often coming home with gashes and bruises, but still smiling from trying new things and pushing her limits.

When she was in elementary school, she often played with the neighborhood kids and got along with almost everyone, regardless of age or background. As she got older, she found that she was able to keep up with most of the boys when a lot of her girlfriends were losing interest in going to the skatepark or hitting the local mountain trails. She was welcomed by the group that met frequently to film each other, and she enjoyed the support and camaraderie they shared.

As she became more involved in the action sport culture, she was often the subject of teasing by classmates outside of her group, but it was more because she was the only girl in her grade who was into action sports on that level, and less about who she was; overall, she was well-liked and friendly with everyone. There were times when girls in her high school could be nasty, and although it was often hurtful, she realized their treatment of her was mostly out of jealousy because she spent a lot of time with the guys on whom they had crushes. Also, her sense of style was heavily influenced by the southern California action sport industry greats she looked up to, and was therefore very different from most of the girls in her school.

Sonya played soccer on the high school team and did well in school. Her mother was very supportive of her endeavors in action sports after school and on the weekends as long as she kept up her grades. By junior year, she had enjoyed enough success at some local action sport contests that she decided to quit soccer and pursue a life on wheels. Her mother hired a private coach and she started building on her bag of tricks. She found

DOI: 10.4324/9781003089629-21

that she was particularly good at the more technical aspects of her sport, so while she continued to work on expanding those, her coach helped her with going bigger and getting more air to impress the judges in her runs.

Action sport culture became more and more important to Sonya, and she started connecting with more athletes in the industry via Instagram and other social media platforms, sharing videos and learning new tricks. Her clips were getting noticed by some of the greats in the industry and she heard that she was being talked about as a potential up-and-coming professional. She started traveling to the West Coast more frequently for contests and practice sessions with different coaches and friends at various training facilities in California. She was able to get small sponsorship items from her local shop back home, but after she won a national contest, she was approached by some bigger companies and her mother decided to help her find an agent. At this point in her career, her mother recognized the potential for her to become a big name in the action sport industry, and while her mother insisted she complete high school with good grades, she saw that Sonya had an opportunity to be successful now, so she did not encourage Sonya to pursue school beyond her high school graduation. Sonya was more convinced than ever that she wanted to be a professional action sport athlete and travel the world doing what she loved most, so she was happy her mother stopped pushing her to go to college like she had in the past.

Unfortunately, before Sonya could sign a contract with any of the big sponsors interested in her, she crashed while practicing at home trying to push one of the big air elements she had been working on for the upcoming contest season. Among the multiple injuries she suffered in the crash was a partially torn ligament that needed surgical repair. Sonya's mother was very supportive and helped her through her recovery, but the sponsors stopped talking to her agent and she had to sit out most of the contest season while she rested, rehabbed, and got back in shape. Her high school graduation was attended on crutches, and as a lot of her classmates talked about their post-grad adventures, Sonya found herself feeling pretty isolated, especially not being able to see all her industry friends on the West Coast who she felt were more like her. The lack of attention on social media and the discontinued communication with sponsors made her recovery even more difficult.

When she was healthy enough to get back to her sport, she decided she needed to move to the West Coast to be closer to more contests and training facilities specifically geared towards action sports. She shared an apartment with two friends in California and worked two jobs to make ends meet so she could pursue her dream of becoming a professional. She got in with a local training facility where many of the local action sport athletes trained, and she worked hard to get herself fitter than she had ever been. Her trainer was stoked on her recovery and fitness, and believed her body could handle all of the elements she was trying to put together.

The facility offered athletic training, physical therapy, massage, and nutrition counseling, so she really felt like she had a supportive team around her and the foundation she needed to be successful. Even though her mother did not have a lot of money, she helped with financial support for the membership at the training facility so Sonya could prioritize her training. Sonya's mother believed she would realize her dream and checked in with her often, offering consistent verbal encouragement. Sonya felt motivated and capable of achieving her goals.

The weather and scene on the West Coast was definitely more conducive to Sonya's pursuits, but she started to realize how many great athletes there were in her sport. She began comparing herself to other up-and-coming athletes and started to doubt her ability to be successful. She considered taking the more artistic road her sport offered by seeking sponsors that did not support the contest route as much; however, her love of competition was something she had had from a very young age and she recognized that she would miss the thrill of the pressure that came with performing on a large stage.

Her first competition back after her injury was one of the biggest contests of the year. She qualified for it the previous year before her injury, and felt like it was her best chance to show that she was back and stronger than ever. She worried about the money her mother was spending to support her training, and her agent made it clear that she needed a good showing at the upcoming contest to get the conversations going again with the bigger sponsors who had shown interest before her injury.

Sonya arrived early to the contest city to make sure she had time to prepare. She went alone; her mother decided it would be too expensive to come watch her compete in-person, so she planned to record Sonya's events on TV and watch when she got home from work. Sonya arranged to share a hotel room with another competitor she had met on the circuit, but she flew in a day earlier than her to have some time to herself. The first day Sonya was allowed in the venue, she was there all day practicing. She did not feel very comfortable with the elements of the contest arena, so she felt like the more time she spent practicing her tricks on them, the better she would be in competition. However, by the end of day one, she felt like there was no way she could land her entire run at this venue, especially not her big air tricks, which she had not felt comfortable enough to even attempt in practice. She felt like she could not breathe well in the venue, and her chest felt tight. Even though everyone was very friendly and outwardly supportive, she felt like all of the other competitors were wondering why she was even there. There were even some new athletes there who were younger than her and throwing down immaculate runs in practice that had the potential to beat her even on her best day. Her coach was encouraging, but she felt like he had other athletes at the competition who would perform better than her. She started to feel that she was not good enough to be there, that she did not want to be on camera doing a run she did not feel capable of completing, and that there was no chance she was ever going to get sponsored.

You meet Sonya at the contest in the training room early in the week during designated warm-ups and before qualifiers. The company that puts on the contest hired you via the medical team, which included athletic trainers, physical therapists, and medical doctors. You were the first mental performance coach they hired for the event, and your role was to be available to the athletes on an as-needed basis as either determined by the athlete or by the medical specialists. Event staff members from the medical team, as well as athlete hospitality, felt that the athletes would be best served if you continued to float around the venue, spending time at as many warm-ups and events as possible, and checking in with the staff at the athlete lounge and medical training room to see if any athletes were interested in your services. Athlete and staff education about your background and what you do was a priority, so you used your networking skills to get to know more of the staff and athletes. You were given a walkie-talkie by the medical staff to be on call for any athletes looking to talk to you.

You heard through one of the staff on your walkie that Sonya may want your assistance because she had a "rough warm-up" and was headed to the medical training room. You went there to meet her and when you arrived you saw that she was very upset. She was getting her body worked on and talking with the staff who she had met at previous events. You were introduced to Sonya by one of the physical therapists and, after chatting, you realized the pressure she was under for her upcoming events. She agreed to talk more with you to see if it would help, but mentions her mother "doesn't believe in that stuff."

Thought Questions

Case Conceptualization

- What information do you need prior to beginning work with Sonya in order to enhance your effectiveness?
- What key aspects of Sonya's past are helpful to understanding her current circumstances?
- If Sonya is healthy and capable, what specific factors might be playing a part in her current struggles?
- Developmentally, what is important to consider as Sonya is navigating the transition from high school to potential professional status as an athlete?

Ethical Considerations

- Do the facts in this case suggest any "red flags?" That is, do you notice the need to consider clinical referral, limits of your consulting competency, competing stakeholders, etc.?
- Even though Sonya is an adult, the group who is putting on the action sports contest will be paying for your services. What elements do you

need to consider when working with Sonya with regards to what she wants, what the contest sponsors/media want to see, and what is safe for Sonya?

Diversity Awareness/Reflection

- What are your initial thoughts regarding the culture involved in this case? These may be related to the event or Sonya.
- What contextual knowledge of the action sports culture do you have to help you in this setting? What contextual pieces are missing that you would like to know more about?
- Is gender an important aspect of diversity in this case? Do you feel as though Sonya being a female in a male-dominated culture has an impact on her overall success in the sport?
- How might Sonya being from a small town impact her as she is acclimating to life as a potential professional athlete in a different part of the country?

Intervention Planning and Evaluation

- Identify the potential issues that may be affecting Sonya (these may be personal and/or performance related). From your list, prioritize the top three issues that you believe are having the most immediate impact.
- What might be some deep-seeded beliefs that Sonya holds that could be influencing her mental and emotional reactions?
- What impact do you think Sonya's mother and coaches have on her? How might they be helping and hurting her efforts?
- How is Sonya's injury and subsequent recovery playing a role in her current state of mind going into this competition?
- What are the sources of Sonya's pressure and what strategies would you incorporate to help her through them?
- Since you are working with her on-site, how would your short-term interventions compare with the long-term interventions, assuming you continue to work with her after the event?

18 Master's Athletes

"The Timid Triathlete"

Jen Schumacher

UNITED STATES

Kathi, a woman in her late-40s, approaches you after a recent talk you gave on the mental game at a local triathlon club. She asks some additional questions and you exchange contact information. She expresses experiencing frustration with her training and performance at races, and asks to set up a call to discuss working with you. You offer her a business card and schedule a time the following week that is mutually agreeable.

During your initial consultation with Kathi, you learn that, while she has been competing in shorter triathlon distances for the better part of a decade, she is a relative newcomer to the Ironman distance, and has been having difficulty with anxiety around the swim and course completion. Ironman triathlons consist of a 2.4-mile swim, a 112-mile bike, and a 26.2-mile run, completed consecutively in that order. The average time to complete an Ironman is roughly 12 hours and 30 minutes, and most courses have a cutoff time of around 16 or 17 hours. While the distance and duration are grueling, arguably the most challenging portion of the event for most athletes is the swim, which often starts "en masse," with many hundreds of athletes in wetsuits running into the water to begin swimming at the same time. This mass start is often chaotic and overwhelming with high levels of contact – often incidental – with other swimmers fighting to find their place in the pack.

Kathi shares that in some of her recent Ironman attempts, the swim has been a miserable experience. Growing up a runner (track and cross country in high school, continued running for fitness throughout adulthood) made picking up cycling a bit easier, but swimming has never been a natural strength. Although she swims regularly, she has never felt confident in the first leg of any triathlon, and throughout the swim she regularly thinks, "I just need to get to my bike." Prior to the start of races, Kathi experiences tremendous anxiety to the point where her wetsuit feels constricting around her chest, and the feeling often persists throughout the swim, especially if she is in a large pack with a lot of contact or in cold water. Before starting any race, her mind immediately goes to past experiences where she has been inadvertently kicked in the chest, slapped over the head by an arm stroke, had goggles pulled off, and even been swum on top of and pushed underwater. Often the contact persists throughout the

DOI: 10.4324/9781003089629-22

entire 2.4-miles because she is not the fastest swimmer, so she can never get out of the pack and away from the contact. When this happens, she feels stuck and becomes even more anxious. She often thinks about quitting during the swim, and actually has on more than one occasion.

If Kathi does "manage to get through the swim," she feels completely depleted of energy yet still has most of the race ahead of her, and occasionally this leads her to second-guess her abilities on the bike, even though objectively she knows she is a strong cyclist. Her doubts about her energy and endurance cause her to back off the pace and put her in jeopardy of missing the cutoff points. In some cases, she will talk herself out of the bike, citing that it is unsafe to be riding when this fatigued, and she might as well quit now if she's going to miss the cutoff anyway. Ending her race during the bike leg seems to happen more frequently when the conditions are challenging, even though she reports training under similar conditions to simulate what to expect on race day. Each time Kathi does not finish (DNFs), she loses confidence in her ability to race the Ironman distance and feels as though she is building a bad habit of quitting. In fact, often when she is contemplating quitting during a race, she will think of how relieved she felt the last time after quitting – each time it gets easier and easier to quit.

Kathi's training group has been very supportive, but she feels the weight of being one of the only women in the group and the only person to DNF on her own (others have been pulled off the course by officials, but no one else has DNF'd of their own accord). Although she is used to being one of few women in any given situation, she is not used to being known as a "quitter," and the idea that others may have that perception of her causes a lot of angst. She is persistent in training and dedicates several hours every day to training and recovery. Despite putting in more distance than many of her training partners, she never really feels prepared enough, and always feels she could do more. She trusts her coach but regularly wonders if she is swimming, cycling, and running enough, and will often supplement her program with an additional workout later in the day. However, when she shows up to races, she still focuses on how others look fitter, faster, and better prepared than her. Kathi mentions being aware that this is an unproductive habit she wants to let go of, and knows she needs to focus more on herself and trusting her training, but she doesn't really know how to do that. Worse yet, triathlon is not the only area in which Kathi struggles with doubting herself and her abilities.

Outside her sport, Kathi is a software engineer who was working for a technology company that recently downsized, and as one of the less experienced engineers, she was let go about six months ago. Since then, she has had dozens of interviews in front of nearly all-male panels. She explains that in the world of coding, where the importance of soft skills pale in comparison to intellectual ability, interviews are less of a conversation and more of a performance – you are given problems to solve in real-time in front of your evaluators. She expresses feeling as though she is exposed on stage in front of a room full of smarter people evaluating her intellectual flaws and shortcomings. This provides an even more intense feeling of anxiety than the Ironman swim – her palms get sweaty, she can feel her heart racing, and she begins to speak much more rapidly and in a higher pitch, which she is acutely aware conveys self-doubt and reinforces the perception that she is a less-qualified older woman. She explains that in software engineer interviews, candidates are expected to verbally articulate what they are doing, and why, throughout the problem-solving process. Sometimes during the interview, Kathi thinks about whether she'll be able to solve the problem and what might happen if she doesn't get this job; this can cause her to "freeze," and she cannot find the solution. She often leaves these interviews feeling discouraged and generally worries that no young tech CEOs are looking for a middle-aged woman who "clearly cannot handle pressure."

Being out of work, regularly going on nerve-wracking job interviews, and underperforming in her sport has placed added stress on her marriage, so she and her husband are going to marriage counseling. They have no children and he is employed, so fortunately finances are not too much of a burden … yet (she wonders how much longer they can keep living in their expensive city on one income). Kathi's insistence on continuing to pursue Ironmans and doubling down on her efforts there, even with everything else going on, has been yet another source of tension in their marriage. However, Kathi feels strongly that working with you on getting her confidence back in her training and racing will transfer into other aspects of her life and help with her professional and personal life challenges. In the past, when she was successfully completing shorter triathlons, she relished in the feeling of accomplishing her athletic goals and felt strong, capable, and unstoppable. It was this success that led her to pursue the Ironman distance, and over the last two years of not being able to achieve this goal she feels like an entirely different athlete. She knew Ironman would be much more challenging than the sprint, Olympic, and half-Ironmans she had successfully completed, but she had no idea it would be this arduous of a journey. She can't bear to give up on the dream of completing her first Ironman, but can't seem to keep that in mind during the race itself in order to go on.

Thought Questions

Case Conceptualization

- What additional information do you need prior to beginning to work with Kathi in order to enhance your effectiveness?
- What aspects of this initial consultation with Kathi do you feel are relevant to building a program to help her enhance performance?

- What do you believe is holding Kathi back athletically and professionally? Do you believe these challenges are related?

Ethical Considerations

- Did anything Kathi discussed raise any "red flags" that might point to the need to consider a clinical referral or limits of your consulting competency?
- Are there any concerns or discomfort about accepting compensation from someone out of work or what might happen if Kathi can no longer afford your services? How might you navigate that potential situation?

Diversity Awareness/Reflection

- What are your initial thoughts regarding the culture involved in this case? These may be related to triathlons and Ironmans, or to Kathi.
- What contextual knowledge of triathlon culture do you have to help you in this setting? What contextual pieces are missing that you would like to know more about?
- What are your initial thoughts regarding Kathi's comments about being the only or one of few women in the room, both athletically and professionally? How might this contribute to the challenges she is experiencing?

Intervention Planning and Evaluation

- Identify the potential issues that may be affecting Kathi (these may be personal and/or performance related). From your list, prioritize the top three issues that you believe are having the most immediate impact and how might you address these.
- What skills does Kathi already possess? How could you capitalize on mental strengths while also developing other areas of performance?
- Kathi seems to be focusing on completing an Ironman and getting a job as measures of outcome success. How else might you conceptualize the evaluation of progress in your work with Kathi?

19 Forced Immigrants

"I Just Want to Play"

Thierry R. F. Middleton

CANADA

At the end of 2019 there were 33.8 million people who had fled their country in search of safety. Each of these individuals has their own story of persecution. However, before delving into the case presented here, there are a few general terms that need to be defined. An *asylum seeker* is an individual who has applied for asylum in the country they fled to in search of safety (UNHCR, 2020). A *refugee* is an individual who has been granted protection by the country in which they sought asylum by going through the process of proving they have a "well-founded fear of being persecuted for reasons of race, religion, nationality, membership of a particular social group or political opinion, is outside the country of his (or her) nationality and is unable, or owing to such fear, is unwilling to avail himself (or herself) of the protection of that country" (United Nations Refugee Agency, 1951, p. 14). A *resettled refugee* is an individual who has been resettled in a third country after having received refugee status from the initial country in which they sought asylum (UNHCR, 2020). For the sake of this case, the term *forced immigrants* is used to refer broadly to asylum seekers, refugees, and resettled refugees.

Syrians continue to be the largest forcibly displaced population worldwide (UNHCR, 2020). Operation Syrian Refugees was a Canadian government program that resettled 25,000 Syrian refugees from November 2015 to February 2016 (GOC, 2019). Canada also resettled the highest number of refugees in 2019. While resettlement is often facilitated by the United Nations High Commissioner for Refugees (UNHCR), 58% of arrivals in Canada were facilitated through private sponsorship groups (UNHCR, 2020). Resettled refugees in Canada receive one year of resettlement assistance from the government and/or their sponsorship group (GOC, 2017). For resettled refugee youth, the Canadian government has provided funding to sport and physical activity programs that help ease the transition of youth into their new community (GOC, 2019).

....

The city in which you live and work has recently welcomed 20 Syrian refugee families. Prior to fleeing Syria, many of these families were

DOI: 10.4324/9781003089629-23

economically stable and lived amongst their extended families. Some families made the decision to leave Syria when unrest began to break out, while other families waited in Syria until it became too dangerous to leave their homes. Dependent on where the families lived in Syria, they either fled to Jordan, Lebanon, and/or Turkey. As of 2019, these three countries hosted the largest number of Syrian refugees in proportion to their population (i.e., 1 in 7 people in Lebanon, 1 in 15 people in Jordan, and 1 in 23 people in Turkey). Before coming to Canada, these families spent an average of two years in the country in which they first sought asylum. After a lengthy application process, they have been paired with a private sponsorship group and/or sponsored by the Canadian government to resettle in Canada. Once the families landed in Canada, they received permanent residency. Receiving permanent residency afforded the families a number of rights that many refugees do not have while navigating the asylum-seeking process, such as full access to healthcare and employment opportunities. Additionally, they can feel more comfortable knowing that they will not be forced to leave their new home. Many of the families also benefitted from having a private sponsorship group, often consisting of multiple families, who volunteered to help their resettlement. These sponsor groups helped families register their children for school, find jobs, and navigate the everyday challenges of living in a new, unknown community, including becoming involved in sport and physical activity programs. Your city is not an urban center, but rather one that could be best characterized as a conglomeration of a small city and several small towns spread out over a vast distance. As such, there are relatively few ethnic enclaves, meaning that the Syrian youth who wish to meet new friends must interact with culturally dissimilar youth to do so.

The number one priority of many families is to ensure that their children are enrolled in school and focused on succeeding academically. School is also often the first place where resettled refugee youth begin to interact with youth living in their host community. As such, school is the context in which resettled refugee youth are introduced to how youth in their host community engage in sport and physical activity. For younger youth, this occurs primarily in physical education classes and on the playground. For older youth, this occurs primarily in physical education classes and through any sports teams they may be fortunate enough to join.

Your city also boasts a number of community organizations that have developed additional support programs for forced migrants. One organization that has committed to helping forced immigrant families integrate into the community is the local recreation center. The recreation center consists of a pool, basketball courts, squash courts, a running track, and a fully equipped gym. Management staff of the recreation center believe that sport and physical activity are one way of integrating forced immigrant families into their community, while also providing them with the opportunity to improve and/or maintain their physical and mental well-being. Wishing to do their part to support forced immigrant families, management offered resettled Syrian families a free one-year membership to their recreation facility. Management have indicated that Syrian families initially seemed to use their membership; however, their involvement in sport and physical activity programs did not last long and no families chose to pay for a membership once their first (free) year ended.

On the board of governors of the recreation center are two principles of a local elementary and high school. Teachers working at these schools have approached the principles and spoken about conversations they have had with some of the male Syrian youth in their classes. Their main concern is that the Syrian youth have indicated they are struggling to become involved in sport and activity programs outside of the school setting. The teachers indicate the sadness that emanates from the Syrian male youth when they speak about not being able to play sport outside of school. Additionally, for many of the male youth in high school, there are limited opportunities to become involved in sport and physical activity. Reasons for limited involvement vary from lack of willingness to try other sports when their preferred sport, which for many youth is soccer, is not available to not having the skill set needed to join a school team. Many of these male youth feel self-conscious about their ability to participate in physical education classes that consist of sports and physical activities they have never encountered. Participating in these activities alongside host community youth who have grown up learning these activities makes them feel unable to fully participate.

When asked about their use of the recreation center membership pass provided to them for one year, many of the youth explained that they were only beginning to explore the benefits of the membership when their complementary year ended. For many, the first year of living in a new city was challenging as they learned how to move around the vast geographical space either by bus or by receiving help from their sponsor groups. A few of the youth have expressed this barrier becoming easier to navigate after their families were able to purchase a vehicle. The youth who did use their membership primarily speak about their use of the weight room and the swimming pool. However, for many, their experiences in swimming lessons were less than ideal as they were often placed in swimming lessons with younger children due to the need for them to develop basic swimming skills. Additionally, many of the youth felt that, while sport and physical activity served to improve their English language skills, paying attention to verbal and non-verbal communication took the majority of their energy and attention, which made paying attention to skill development a harder undertaking.

Many male Syrian youth also eagerly speak about the opportunity to join a competitive soccer team. Soccer is a sport they grew up playing and many wish to train just like the professionals they watch on television. Some of the youth have been lucky enough to have the opportunity to join a team, primarily due to support from their sponsoring group. Other youth have primarily relied on playing soccer with an Arabic-speaking group that gets together to play soccer once a week. Part of the enjoyment in playing with this group is the opportunity to communicate in the mother-tongue language of all participants. This group is mainly comprised of adult men (i.e., 20–60 years old) and so does not provide younger teenagers any interaction with youth their own age. For the youth who have not had an opportunity to join a team yet, any opportunity to play is relished; however, they do wish to be able to play with kids their own age and use soccer as an opportunity to get better at speaking English.

The board of governors of the recreation center has tasked management staff with developing a plan for how support of forced immigrant youth can be improved through sustaining their involvement in sport and physical activity programming. Correspondingly, management staff have approached you, as an expert in the development of, and working within, youth sport and physical activity programs, to help them as an outside consultant. While they know who they would like to serve, they also highlight that they do not ask members to indicate their residency/immigration status. As such, there is no easy way to track how many forced immigrant families may be using the recreation center at this current time.

Thought Questions

Case Conceptualization

- What information do you need about this organization prior to beginning to work their management staff in order to enhance your effectiveness?
- What information do you need about the forced immigrant population in your community prior to beginning to work with this organization?
- What information do you need about broader political, historical, and/or societal factors prior to beginning to work with this organization?

Ethical Considerations

- Do the facts in this case suggest any "red flags?" That is, do you notice the need to consider clinical referral, limits of your consulting competency, competing stakeholders, etc.?
- As a non-forced immigrant what non-clinical factors do you need to consider when working with a vulnerable population?

Diversity Awareness/Reflection

- What are your initial thoughts regarding the culture and cultural differences involved in this case? These may be related to the performance arena or the individual performer.

- How do you plan to learn more about the cultural backgrounds of the youth you will be asked to develop programs for?
- How will you account for divergences in the values and desires held by different stakeholders?

Intervention Planning and Evaluation

- What intrapersonal barriers exist that limit forced immigrant youths' involvement in host community sport and physical activity programs?
- What interpersonal barriers exist that limit forced immigrant youths' involvement in host community sport and physical activity programs?
- What structural barriers exist that limit forced immigrant youths' involvement in host community sport and physical activity programs?
- Identify the potential issues that may be affecting the performer (these may be personal and/or performance-related). From your list, prioritize the top three issues that you believe are having the most immediate impact.

References

Government of Canada (GOC). (2017). *Syrian refugee integration – one year after arrival*. Retrieved from https://www.canada.ca/en/immigration-refugees-citizenship/services/refugees/welcome-syrian-refugees/integration.html

Government of Canada (GOC). (2019). *Syrian refugee resettlement initiative – Looking to the future*. Retrieved from https://www.canada.ca/en/immigration-refugees-citizenship/services/refugees/welcome-syrian-refugees/looking-future.html

United Nations High Commissioner for Refugees (UNHCR). (2020). *Global trends: Forced displacement in 2019*. Retrieved from https://www.unhcr.org/5ee200e37.pdf

United Nations Refugee Agency. (1951). *Convention and protocol relating to the status of refugees*. Geneva: UNHCR Communications and Public Information Service. Retrieved from http://www.unhcr.org/protect/PROTECTION/3b66c2aa10.pdf

20 Athletes with Disabilities

"Just Get Over It"

Andrea Faull-Brown

United Kingdom

Jane is a 32-year-old highly experienced female wheelchair basketball player with over 12 years of playing experience since acquiring her disability during a skiing accident. The accident resulted in a spinal cord injury causing her to have to use a wheelchair for daily mobility as well as for playing wheelchair sports including wheelchair tennis and basketball. Jane still skis (sit skiing) despite the injury, which is a complete lesion at T6, meaning she has limited function from below the sixth thoracic vertebrae in her back; as a result, she is classified as a "1.5 point" player in wheelchair basketball. In wheelchair basketball, only five players (totaling 14 points) are allowed on the court at any one time, and for this reason she is considered a key player on the team. Jane has played wheelchair basketball at the Paralympic Games and World Championships standard as well as playing at Club level, so she has experience dealing with large crowds and high-pressure environments but also has experience with and an understanding of the demands of Club level competition. Jane works incredibly hard at her sport, and that was recognized by most of the squad; she always trains hard and thinks little of driving over 150 miles to training several times per week all whilst holding down a full-time job.

Jane is not the current captain of the team (although has been in the past) and genuinely embraces this, recognizing that there are younger, "up and coming" players who need the opportunity to hold that position. However, given her age and experience, her views are valued and often sought after when players are asked for input or when there are breaks in play during training and competition. For this reason, the coach also values her input and knowledge but only appears to make use of select information offered up by Jane. This is more a reflection of the coach than Jane, but this inconsistency in behavior by the coach can, at times, cause confusion, frustration, and mixed messages. Jane generally has a measured way of communicating with her teammates but can, at times, be somewhat emotional, which can result in others thinking she over-reacts to issues. Jane can control her emotions, but when there are things that she doesn't agree with or there are decisions or times when she and her playing time are affected, her emotions can be a challenge for her to manage appropriately.

DOI: 10.4324/9781003089629-24

Jane captained her National side during a major World level competition (which was around two to three years ago) and it was around that same time that she first noticed the decrease in her performance with her free throw shooting ability. Free throw shooting is a key part of wheelchair basketball and Jane has always been considered a good free throw-taker, being known to seek out opportunities to take them in games as well as capitalize on them in end-game scenarios. But this has recently changed, and Jane reported that it was becoming so bad that she would avoid situations in a game which might result in her being fouled and having to go to the line to take a shot. When faced with taking a free throw, Jane resorts to an old free-throw routine that she had developed around four years ago and admitted to making use of rather inconsistently. She moves to the line, positions her chair, adjusts the ball until it feels "right" in her hands, takes a deep breath, and then shoots. Often, the shot fails to get the height and power required to even hit the rim of the basket and veers off course dramatically, so it is not on target or powerful enough to be considered a valid shot. This is unusual as Jane is physically strong and accurate enough to make the shot. She admits to having both physiological and psychological issues at the line including a range of negative thoughts like feeling distracted and worried about messing up the shot, and uncontrollable heart rate and breathing.

When asked about what, if anything, had been done to date to try to resolve the issue with her free throws, she reported some pressure training (the coach makes them watch her take shots and then be punished as a group for her mistakes which involved bouts of sprints) as well as watching some video of her current technique to try and work out what needs to change. You witnessed the punishment first-hand during a training session where pressure training was being used and you could see that the effects of it were impacting not only Jane but also the rest of the squad; it was very uncomfortable to watch. When asked more about the pressure training, Jane confirmed your suspicions that the punishment was causing tension in the squad. Ultimately, however, Jane had gotten to the point where she was no longer worried about making the squad *push* (sprint). She knew that they were finally starting to understand that she was struggling with the issue, but equally team members were starting to lose patience and would tell her to "sort it out" or "just get over it." This was particularly challenging given that you are offering support to both Jane as an individual and to the whole team, so the tension and conflict that was arising was becoming an issue to manage on its own aside from the individual needs of Jane. The team punishment as a result of the pressure training was used so frequently that Jane admitted this was no longer effective and its purpose (which was to induce some anxiety in the hopes that this would help in real-game scenarios) was no longer working, making any type of pressure difficult to simulate in practice as it had been overused/misused to date. You agreed that it wasn't effective and made a mental note to talk to the coach about it.

Although the coach was initially supportive of Jane receiving some support on the issue and stated that he was happy to do some sessions in training that might help your work with Jane, it was clear that the coach was also starting to lose patience with Jane and her issue. As a result, it was starting to impact her court time, as the coach was becoming increasingly unsure of Jane's ability to perform in end-game scenarios, when she would most likely be fouled and sent to the line. Consequently, Jane was being taken off in the final minutes of play, but it was at this very time that her experience was needed on the court. This was adding to Jane's frustration (and that of the other squad members) about the issue. Jane reported that her anxiety during free throws was a combination of cognitive and physiological symptoms. Other aspects of her game were, at the time, reasonably unaffected and she appeared to generally handle the pressure of the rest of a game exceptionally well, so it only seemed to manifest itself negatively in free throw scenarios. However, knowing that the coach was losing patience appeared to be adding to Jane's anxiety, and communication between the coach and Jane was starting to feel strained and difficult when previously that had not been the case. Finally, Jane's partner and friends would frequently watch games and it was clear that when Jane was sent to the line to take a free throw, there was a hushed tension that could be felt not only on the court and on the bench but in the stands, too. Therefore, it was apparent that others were aware of Jane's issue and this was clearly adding to the anxiety that she had around taking free throws.

Thought Questions

Case Conceptualization

- What information do you need prior to beginning to work with this individual/group in order to enhance your effectiveness?
- What contextual knowledge of the wheelchair basketball and Jane's impairment do you have/need to have to help you in this setting? What knowledge do you have around working with athletes with a physical impairment?
- What key aspects of Jane's past are key to understanding her current circumstances or are these unrelated?

Ethical Considerations

- Do the facts in this case suggest any "red flags"? That is, do you notice the need to consider clinical referral, limits of your consulting competency, competing stakeholders, etc.?
- Given that wheelchair basketball is a team sport, what impact could your involvement with Jane have on the rest of the team and how could you best manage this?
- How can you mitigate from the others feeling that Jane is the "favorite"?

Diversity Awareness and Reflection

- What are your initial thoughts regarding the culture(s) involved in this case? These may be related to the performance arena or the individual performer.

Intervention Planning and Evaluation

- What challenges might you face when designing an intervention for an athlete with a physical impairment? What psychometrics are available and suitable for use? If there are none available, what adjustments might you be able to make to existing psychometrics (which are mainly aimed at non-disabled athletes)?

Section 5

Ethical Considerations in Consulting

21 Who Is the Client?

"The Team, the Coaches, the Manager, the Owner, the Players ... and the Sport Psychologist"

Matt LaCoste

FRANCE

The Roosters are a professional rugby team with a glorious history. In the past, the Roosters were one of the best teams in their country's Elite championship league and qualified for the European Cup several years ago. However, for a few years now, the team's results have been up and down. The team has lost its status, and its position decreased step by step.

The owner changed coaches several times, he renewed players, but results and team performances still did not match the board's expectations. This year, the owner, in an attempt to bolster the team's leadership, hired two new assistant coaches who were famous former Roosters' players. Both of them have been captain of the team, and they have played within the group for many years. They know the club and all the former players. The preparation for this new season started a couple of weeks ago.

Through a mutual relationship, Dr. Ment, a sport psychologist, met with the new Rooster's head coach. After an informal conversation about sports, performance, and mental aspects of elite players, the head coach seems interested in introducing mental training into the professional team's weekly practice. It would be the first time the club would use this aspect of sport performance.

The pre-season preparation begins with daily training, and the coaches already scheduled some friendly games. Dr. Ment shows up regularly to the training center, and sometimes he has an informal talk with the head coach. The team leaves for one week of intensive training at altitude, isolated from media and players' families. Because the terms of the contract have not been discussed yet, Dr. Ment did not participate in this training camp.

Dr. Ment is happy with this new challenge. He starts to come and watch the team during training. He tries to develop his evaluation of the group's dynamic, potential leadership, and the players' personalities. Dr. Ment also observes coaches-players relationships.

The season begins. The first official game at home was a defeat. The team lost five successive games and started the season in the bottom two

DOI: 10.4324/9781003089629-26

(drop zone of the championship). Dr. Ment stays in touch with the team; he followed almost all games at home and even away. He is still coming to watch 2 or 3 training sessions each week and is usually invited to participate in the after-game party with players. In this way, Dr. Ment develops bonds and relationships with all players. Indeed, after the games, two types of post-match gatherings are organized. In one case, it is required for players to meet the team's sponsors who support or fund the team; in the other case, it is only for players to share a drink and spend a friendly time altogether.

After the fifth defeat, the manager of the Roosters calls Dr. Ment for an interview. Dr. Ment prepares the meeting and presents his view about mental training to the manager. Here, the manager is the sport director; he oversees the entire club's performance, supervises the coaches, is the primary decision maker about the recruitment, and is a member of the club's board of directors. During this meeting, they discuss how mental training can improve individuals as team performances and what and how he can help the team and the players. The manager seems to agree on introducing mental training into the players' weekly preparation. However, a disagreement arises when the discussion turns about the budget. Dr. Ment has planned to follow the team very closely. He wanted to be part of the coach's group to be able to have contact with players and coaches almost daily; during training and games. Thus, Dr. Ment asked for a salary matching with the time and energy he will invest in that role. However, the manager had other plans regarding the Rooster's championship position, and he planned to invest in new players to reinforce the group. Thus, the manager offers a role "in-between" to Dr. Ment. The club will pay only one day per week and coming to the game (without access to the locker room) is not mandated. The manager and Dr. Ment will rediscuss this contract at the half of the season.

Already in contact with most of the players and already in support of the head coach, Dr. Ment accepts the role with this small capacity for intervention and action. The following week, the coaches organize a meeting with all the players to reintroduce Dr. Ment and clarify his role with the team. Dr. Ment noticed that one of the assistant coaches does not show the same enthusiasm as the head coach. During this first team meeting, Dr. Ment also saw a certain urgency inside the team and worries from coaches regarding the beginning of the season, even if they won the sixth game.

Dr. Ment officially starts his function as a mental trainer for the Roosters Elite Team. He meets individually and collectively with players. He assesses all the players' psychological aspects of performance, self-confidence, motivation, coping strategies, relationship inside the group, and the coaches–player relationship.

Even if the manager or coaches did not mandate it, Dr. Ment comes to see almost all games. He takes the initiative to talk to players who prepare to play during the second half. Dr. Ment watches games in the stadium stands, close to the players not selected or injured. Likewise, Dr. Ment develops a bond with the team Doctor to discuss the injured players. He also continues to join after-game parties with players.

During the week, Dr. Ment has sessions planned with players. He works with injured players to help them through their recovery process. He organizes group sessions with all the forwards and group sessions with players who need to improve a specific mental aspect of their performance. Dr. Ment spends time with the team's main striker to increase his ability to concentrate, visualize the process, and cope with pressure. He also offers individual sessions about personal life with some players who experience sleeping disorders, family or couple issues, and anxiety. However, he has no more contact or meetings with the manager, and he never meets with the owner or the board.

For a few weeks, they stopped the downward spiral of defeat, but the team alternated wins and losses, and they stayed in the bottom half of the league table. Dr. Ment develops a good bond with nearly all players of the team. They cooperate more and more and become more involved in their mental training. Dr. Ment gets invited to player's parties, where coaches are not.

Dr. Ment realizes that he has fewer occasions to talk with the head coach, who seems to be in trouble with the owner and the manager regarding the team classification. Indeed, it is almost half of the season, and the results do not match the pre-season expectations. Moreover, Dr. Ment does not work with the head coach on the psychological aspects of coaching, his role with the team, and the actual situation. Nevertheless, he builds a better relationship with the assistant coach, who appears increasingly interested in Dr. Ment's role and work. He also seems more confident about the benefit of mental training for the team and the performances.

Step by step, the results become a bit better even if it is still hard for the team to perform consistently, and the lack of consistency relegates the team to the bottom of the championship's league. The atmosphere between players is pleasant, but some frustration between players and coaches arises game after game. Dr. Ment maintains his weekly session with the players who feel the need. Injured players and those who play less often are the players who come and talk the most with Dr. Ment. A club-wide focus on short-term results becomes apparent. Dr. Ment observes this growing tension inside the group and lets the coaches know. The head coach gets upset by the information and asks Dr. Ment to focus on that goal. He mentioned that Dr. Ment's interventions with the players should not aggravate the tension even if he understands that venting frustration could be useful for players. Some players also suggest to Dr. Ment that they struggle to see him as fully part of the coaching staff, and some of them do not feel the full support from the coaches towards his role within the group. Disappointed at first, Dr. Ment quickly understands that the pressure for results could be more robust than what he had anticipated. Dr. Ment decides to listen more to the coaches, offers them a space to speak, provides his support, and gives them some coping advice.

Before the mid-season break, the board fired the head coach. A new head coach coming from outside of the club takes the responsibility of the team. The assistant coaches are maintained with the same functions. Dr. Ment sends a report of his work to the new head coach, who gives him positive feedback.

During the break, Dr. Ment meets with the manager, who informs him that the club will not renew his contract for the second half of the season. The manager explains that the decision comes from the board because of the restriction of the budget. The club should anticipate a relegation of the team to the lower level. Indeed, the team rarely rose above the relegation zone during the first half of the season.

Unfortunately, the second part of the season has not been better for the Roosters, and the team experienced relegation at the end of the season.

Thought Questions

Case Conceptualization

- Who is the client of Dr. Ment in this case?
- What does the Sport Psychologist need to know before starting working with the Roosters?
- What components of the club dynamic were important for Dr. Ment's intervention? Why?
- On which psychological elements of the team performance should Dr. Ment focus?
- What strategies and organization could Dr. Ment implement to reach these goals?

Ethical Considerations

- Do the facts, in this case, suggest any "red flags" you could identify as over the limits of your competency?
- How and what elements of the team evaluation can be shared with coaches?
- How and what information obtained during individual sessions can be shared with coaches?

Diversity Awareness/Reflection

- Do you think that perfect knowledge of the sport (rugby rules, rugby culture) was necessary for Dr. Ment? As a Sport Psychologist, what aspects of rugby could influence your work?

- How does the glorious past of the club impact the team and the players? How could this impact be part of mental training? How could these factors be taken into consideration by Dr. Ment?

Intervention Planning and Evaluation

- What does Dr. Ment do well in his approach and his work?
- Identify some mistakes done by Dr. Ment and what improvements he could have achieved during his time with the club and team?
- In this situation, what would you do differently from Dr. Ment? Identify the top three issues that had the most crucial impact on team performance?

22 Consulting versus Counseling

"Kio's Concussion Conundrum"

Megan Byrd and Brandonn Harris

United States

Kio is a 25-year-old Asian male. His family is originally from Niihama, Japan; they moved to Boston, Massachusetts when he was three years old. Kio has played lacrosse since moving to Massachusetts, after he saw his neighbors playing the sport in the cul-de-sac. Growing up, he had opportunities to participate in various organized sports such as baseball, soccer, and swimming, but wanted to follow after his older brother, who played lacrosse at a high level, and now coaches high school lacrosse in Boston. Kio was recruited to play at a private university in New England, where he competed for four years. Currently, he plays for a semi-professional team in Maryland with the hopes of advancing to a team within Major League Lacrosse.

Kio is in his third season with the team and was elected by his teammates as one of three team captains during his second and third seasons. The team employs two Certified Mental Performance Consultants® (CMPC), one male and one female. Both consultants have master's degrees in Community Counseling and PhD's in Sport and Exercise Psychology from a Sport Science department. One is also a National Certified Counselor. Kio's team participates in group sessions once a week, and players have the opportunity to meet with the consultants on an individual basis as needed. Kio was an active participant in the team sessions, but throughout his first two seasons had not sought out individual sessions.

During his first year with the team, Kio performed well at his position and got along well with his teammates and coaches. At the end of his first season, his coaches told him that he was a "stand out" in his position and they believed that, as long as he continued playing this way, he would put himself in a favorable position to advance within semi-professional and professional divisions in the sport. After his first season concluded, he went home and continued to train with his brother to prepare for the upcoming season. He returned to his second pre-season feeling healthy and in better shape than his rookie season. The season started well, with his team going undefeated in the pre-season and winning the first three games of the year.

During the first game of his second season, Kio was making a play on the ball and was cross-checked in the back of his head by a defender who was attempting to secure the ball. Following the hit, he was pulled off of

DOI: 10.4324/9781003089629-27

the field by the athletic trainer for an evaluation. The athletic trainer's examination revealed that Kio had suffered a concussion, the third concussion of his career. Soon after his diagnosis, Kio began the return-to-play concussion protocol which involved a series of cognitive, balance/stability, and visual tests, along with a symptoms checklist (e.g., headache, nausea, dizziness, emotional disruption). Because Kio had suffered from concussions throughout his career, he felt equipped to handle the rehabilitation process given his successful return-to-play experiences in the past. After seven days of being symptom-free, he successfully completed the protocol and was cleared to play.

In the fourth game of the season, three weeks after his concussion diagnosis, Kio's team and their opponent were tied at the end of regulation. Three minutes into the first overtime, while helping to defend his team's net, an opponent running across the field collided with Kio and inadvertently struck the back of Kio's head in the process. Both Kio and the opponent were rendered unconscious briefly, with both teams' medical personnel rushing to aid the players immediately following the collision. Players from both teams were visibly shaken as the two players laid on the ground. After a few minutes, Kio was escorted off of the field and was evaluated for a concussion. Subsequently, after a thorough evaluation, Kio was diagnosed with his second concussion within a three-week period.

Subsequent concussions often lead to a longer recovery process, and for Kio, this was the case. His symptoms were more severe and did not subside for many months. Kio reported experiencing difficulty remembering, concentrating, and sleeping. Although cleared to participate in team sessions by his athletic trainer, Kio sat out during activities that included ball tracking or excessive movement, memorization, or other cognitive demands, as he struggled with these tasks. The CMPCs also noticed changes in Kio's mood, withdrawal from some team activities, and impulsive behaviors. Additionally, team physiologists and athletic trainers reported that Kio was not improving on concussion-protocol tasks. As Kio remained stagnant in his recovery, he sought out a second medical opinion/evaluation and, as a result, was also diagnosed with whiplash; he was advised to implement a different rehabilitation protocol to address his lingering symptoms. Due to the nature of his injury and continued complications, he was unable to return for the remainder of his second season. During this time, the CMPCs reached out to him and voiced concerns to the coaching staff regarding his mood changes and withdrawal from teammates. Although Kio appreciated being checked on, he chose not to engage in individual mental training sessions.

At the conclusion of the season, Kio returned home, and again trained with his brother to prepare for his third season with the team. He also continued to receive medical care related to his lingering symptoms. When his third season began, Kio was elected to continue serving as a captain and was optimistic and excited to compete again. Pre-season training began, and Kio felt he was in the best shape of his career. He reached out to one of the CMPCs expressing an interest in meeting, as he felt he was in a place where it was a "tremendous time to grow and improve in several areas." Shortly after training began, his grandmother passed away and he missed a week of training to go home and attend the funeral. When he returned to the team, he continued to express interest in mental training, but ultimately did not follow up to schedule a time to meet.

Two weeks before the first game of the season, during a training session, Kio planted his leg on the field, which was wet from rain the day before, and twisted his knee. As a precaution, he was carted-off of the practice field and evaluated by his athletic trainer; he was visibly upset. An MRI revealed that Kio had suffered a lateral and medial tibial plateau fracture. This would not require surgery; however, Kio was given a 12-week rehabilitation plan. Not only was his return-to-play for the season in question, but coaches and medical personnel questioned whether he would be able to, or if he would choose to, play again given his extensive injury history over the last two seasons.

The team has weekly staff meetings that include the coaching staff, physiologists, athletic trainer, and CMPCs. In these meetings, the players are discussed regarding their physical and mental recovery, as well as injury status and updates, and general information about the week. During one of these meetings, the coaching staff and athletic trainer discussed concerns regarding Kio and asked the CMPCs to put him on "their radar" due to observations about Kio's mood and general affect. They asked the consultants to reach out to him to see how he was coping during his rehabilitation. At practice, the consultants had been able to speak with Kio as he was doing his exercises on the sidelines. He reported that he "was not doing well" and felt concerned that the coaching staff was rushing him to return to play.

Thought Questions

Case Conceptualization

- What information do you need prior to beginning to work with Kio in order to enhance your effectiveness?
- What carry-over effects might be happening from Kio's previous injuries, as well as the loss of his grandmother?
- Given the information provided, what from Kio's story and experience may be best conceptualized as pertaining to performance enhancement, and what may be more related to mental health concerns?

Ethical Considerations

- Do the facts in this case suggest any "red flags?" That is, do you notice the need to consider clinical referral, limits of your consulting competency, competing stakeholders, etc.?

- How might the consultants' training impact the ethical considerations pertaining to service provision for Kio?

Diversity Awareness/Reflection

- What are your initial thoughts regarding the culture (i.e., sport culture, Kio's Asian culture, professional sport culture) involved in this case? These may be related to the performance arena or the individual performer.
- How might the consultants' characteristics (e.g., gender, ethnicity) influence Kio's reluctance or commitment to working on mental skills?

Intervention Planning and Evaluation

- Identify the potential issues that may be affecting Kio (these may be personal and/or performance related). From your list, prioritize the top three concerns that you believe are having the most immediate impact and what type of support personnel would be best equipped to provide this support.
- Based on your list of top concerns, what potential barriers do you anticipate will arise as you work with Kio?
- If Kio is unsure if he would like to further his work with the mental performance consultants, in what ways could you continue to support him?

23 Tele-Consulting

"The Stagnating, Anxious Swimmer"

Abby Keenan

UNITED STATES

After swimming competitively year-round for the past eight years, Tyler needs a break. He's 16 years old, ending his sophomore year in high school, and can't bring himself to transition from his winter short course championship meets into the upcoming summer long course season. The anxiety, self-doubt, and pressure to perform in order to swim in college are incredibly overwhelming.

At a high-caliber, year-round club in a prominent metro area, Tyler excelled in swimming from ages 8 to 13. The pool felt like a second home – not just because it was right around the corner from his house, but he also had lots of friends there. He loved his coaches, the environment was high-energy, supportive, and fun, and his older sister swam there too. This was one of the few places he could really be himself. As practices became progressively harder over the years, he continued to grow as an athlete, getting faster and stronger.

Around three years ago, he suddenly hit a plateau. Not only did his times stagnate, so did his overall experience in the sport. He felt left behind and left out as his friends slowly all moved up to higher groups. They started to look down on him and taunt him, telling him he'd never be good enough. His mom described this as bullying, wondering if he internalized those experiences. At times he even felt ignored by his coaches, almost like they had given up on him. It finally got so bad that about a year ago, after his freshman year of high school, he decided to leave his club to swim for a competing club in the area. What did he have to lose?

At the new club, he found renewed energy and hope. Surely this change would help him mix up his training, find his confidence again, and get back on his path to swimming in college. Like many other competitive swimmers, he chose not to practice with his high school team in order to prioritize time at his club. After a few months of demanding training, it seemed to work. He was back to dropping time at meets. He also noticed a positive impact on school, achieving A's and B's. He was mostly managing his work even though it was common for him to procrastinate – a lot. With so much invested in his grades and swimming, he started to realize his social life was lacking.

DOI: 10.4324/9781003089629-28

Tyler often felt sad while swiping through social media, seeing friends having fun, and enjoying their lives. He hadn't exactly been making time for friends as much as he thought he should. How did they do it? Swim practice alone was two or more hours a day, six days a week, plus morning practice twice a week, on top of the drive to and from the pool, which was now up to two hours round trip with his mom or dad. He had no extra time between school, practice, homework, studying, and meets. It felt like he could only have one or the other – get good grades and swim fast so that he could swim in college or quit swimming and have a social life. All he kept envisioning was swimming in college; he longed for that team-oriented, fun, challenging environment that would become his home away from home.

In the fall of his sophomore year, Tyler started to build a relationship with his new coach, who was experienced (he had been coaching for over 20 years), respected, and supportive. He had high expectations of Tyler and saw a lot of potential in him, often saying, "You are a better swimmer than you think you are. You have the physical ability, so whatever is holding you back must be mental." Tyler and his coach discussed various goals, such as getting a qualifying time for Sectionals in 100 Back, a State cut in 200 IM and 200 Back (he already qualified for 100 Back), and dropping time in 200 and 500 Free. His coach felt Junior Nationals was even attainable before the end of high school. These goals seemed within reach based on his current best times and training, although according to Tyler's dad, Tyler never wrote down his goals or took intentional steps to work toward them.

Over the next few months during short course season in the fall and winter, Tyler continued to improve but not quite at the rate he anticipated. He became more anxious at meets, putting pressure on himself to reach certain times and trying not to disappoint his coach in the process. Each time he had a bad first race, didn't get the time he wanted, or was beaten by someone he would normally beat, Tyler's confidence dipped and his overthinking increased. Focusing on specific events and trying to drop time only seemed to make things worse, perpetuating the cycle. Interestingly enough, his mom noticed that Tyler was doing quite well in the events he didn't specifically focus on at meets.

That winter, in addition to the ups and downs with swimming, Tyler's family received news that his mom was diagnosed with cancer. Luckily, the doctors caught it early, enabling their close-knit family to act quickly. His mom underwent surgery and treatment halfway through his season and school year. While she was recovering well and her doctors were pleased with her progress, Tyler was experiencing general anxiety in many aspects of his life, so he began working with a therapist.

Additionally, at his last meet of the season in March of his sophomore year, Tyler hurt his back. He took two weeks off from swimming, completed physical therapy, and was cleared to return to the pool. He had never been injured before and thought it was related to overuse. In talking with you about it now, a few months later, he says it feels fine and doesn't make a big deal out of it; however, when he got back in the pool, he immediately noticed how much fitness he had lost. He was falling behind and falling apart when he couldn't keep up. Some days, if he was tired, not feeling well, or knew they'd be working on something he was not good at, he didn't want to go to practice. Meltdowns in the car weren't uncommon. This led to what he described as "panic attacks" during practice, taking feedback personally, and making excuses to get out of the water. His coach initially encouraged him to stay in and keep swimming, even if he had to sit out an interval and then go on the next one. Soon Tyler felt like he was on his coach's "bad side." He was no longer allowed to get out to use the restroom or get his inhaler, but rather was told he had to finish a set. After this became a pattern, his coach expressed concerns about Tyler's ability to endure the upcoming summer long course training. With 23 grueling hours in the pool each week plus dryland training, his coach said that many kids crack and want to quit, even the mentally tough ones. He asked Tyler if he wanted to move down a group or take some additional time off.

After discussing this with his parents, Tyler decided to take another break from swimming, continue working with his therapist, and focus on school. The extra time to do homework, study for finals, and hang out with friends was nice. Although he missed swimming and his teammates, he actually enjoyed the newfound time and freedom at the end of his sophomore year.

He also made notable progress in therapy; he learned a lot about himself, his general anxiety, and various coping strategies. When appropriate, his therapist tried to relate strategies to swimming, such as using a stress ball, visualization, and positive self-talk. Additionally, he started taking a low dose of an SSRI, helping him develop more control, energy, and feelings of well-being. After several weeks, Tyler's therapist asked him if he wanted to try getting back into the water gradually and work with you, as a mental performance consultant, to prepare for his transition back to the club.

While he took time off from swimming, Tyler had been working out with his dad at a local gym. In late spring, he started swimming there for 45–60 minutes after workouts. He always had a lane to himself, which he liked. He was able to monitor how he was feeling and go at his own pace. Sometimes he would get a little anxious or worry about having a panic attack. At those times, he tried to tell himself, "this will pass," "keep going," "it will be fine," and "it will eventually go away;" these were thoughts suggested by his therapist. To calm down, he would try to imagine being somewhere else. Unfortunately, he didn't experience much success with those strategies as the same anxious feelings kept creeping up. While he wasn't doing anything in particular to prepare for each swim, his main goal was to get in the water and just keep going.

With a few weeks of swimming at the gym under his belt, and long course season already underway, Tyler began thinking about what he wanted out of swimming moving forward. The NCAA recently adjusted its college recruiting rules, shifting the opening of recruiting communication from July 1 after an athlete's junior year to June 15 after an athlete's sophomore year, which was right around the corner. Tyler believed he had a good coach; however, it bothered him that his coach didn't seem to understand his commitment to school or take his mental health seriously. He considered changing clubs again, but ultimately decided against it because he was moving to a smaller, private school in the fall for his junior year and wanted to have at least one constant during that transition. After much discussion around his needs and goals with his parents, his mom searched for "mental coaching" online and found your website.

After connecting with Tyler's mom and dad as well as Tyler himself, they agreed to begin working with you. Tyler is looking for sport-specific support and to build mental skills that will help him return to his club in a few weeks, get back in shape, navigate anxiety and self-doubt in practice and at meets, and work toward achieving his goals in his best events. His parents seem very supportive. They emphasized that while finances are a little tighter than normal, they are willing to do anything to help him get back on track with swimming because at one time they saw how happy it made him.

Thought Questions

Case Conceptualization

- What additional information do you need prior to beginning to work with Tyler in order to enhance your effectiveness?
- Developmentally, what is important to consider about Tyler's age in order to understand his current experience, struggles, and abilities?

Ethical Considerations

- Do the facts in this case suggest any "red flags?" That is, do you notice the need to consider any clinical referrals, limits of your consulting competency, competing stakeholders, etc.?
- Tyler is under 18 years old, so his parents will be paying for your services. Who is your client? How will you involve his parents while appropriately navigating confidentiality?
- Tyler will be working with both you and his therapist concurrently. What do you need to consider in terms of competencies, boundaries, and confidentiality?
- If your work with Tyler will be online, what ethical principles and standards should you be aware of and follow?

Diversity Awareness/Reflection

- What are your initial thoughts regarding the culture involved in this case? These may be related to the performance arena or the individual athlete.
- What knowledge and/or experience do you have related to swimming that could support your work with Tyler? What do you need to learn more about to be effective in your approach? How will you go about that process?
- What assumptions about Tyler's race, gender, and sexual orientation did you make in reading this case? What personal biases would be helpful to explore?

Intervention Planning and Evaluation

- Make a list of the issues and opportunities for growth presented by Tyler and his parents. What mental skills and strategies would you prioritize if you had five one-hour sessions to work with Tyler? Why?
- What technology, processes, and strategies will you use in working with Tyler and evaluating your work? How might you change your approach if you were working with him online rather than in-person?

24 Consulting Boundaries

"The Burned-Out Consultant and the Importance of Self-Care"

Sebastian Brueckner and Göran Kenttä

GERMANY AND SWEDEN

Mike at 49 years old is an experienced professional in the field of sport psychology. He holds a dual position as a part-time mental performance consultant and teaching sport psychology as adjunct faculty at a local college. He grew up and currently lives in Europe, though not in his home country, after returning from graduate studies in the United States. His professional career is supported by a strong network of national and international collaborations and he is certified nationally and internationally as a mental performance consultant. In line with his professional philosophy, grounded in a scientist-practitioner approach to applied sport psychology, he has engaged in a number of applied research projects. These projects have helped him to develop and inform his applied practice at the high-performance training facility where he consults. Since graduating, Mike has continually built his network through conference presentations, publications in refereed journals, and more recently invited keynotes. Through his education and collaborations, he advanced his expertise in consulting techniques and counseling approaches, now offering services that focus on clients developing in a holistic and sustainable way, both performance- and personality-wise. As such, Mike's career is grounded in a strong value system. Committed action and a client-centered approach, a strong work ethic, advocacy for those with less power and a holistic, humanistic perspective are core pillars for his work. Trust, empathy, openness, and authenticity are core values for Mike.

A strong work ethic and, at times, perfectionistic tendencies are rooted in his past as an elite athlete. He competed as a distance runner for 13 years, qualifying for Nationals and Olympic trials. His interest in and the decision to pursue a career in applied sport psychology was, at least partially, rooted in his experiences as an elite athlete. In trying to excel, to reach his full potential in training and competition, he realized the importance of mental training and its positive effects on him – not only from a performance enhancement perspective. As a high achiever, he sometimes struggled with, what he now knows as, mental health issues in trying to balance athletic demands with pursuing his bachelor's degree in sport

DOI: 10.4324/9781003089629-29

science. Rumination, insomnia, excessive worry, and social withdrawal were symptoms he experienced. At that time and place, he did not have the resources to help himself and self-care was a low priority for him.

Upon graduation, Mike ended his running career and decided to pursue graduate studies in the United States in applied sport psychology – an emerging professional field internationally. Over the years, he managed to build a successful professional career. His successful consulting was evidenced by funded research projects, high demand for his workshops, positive feedback in his academic courses, as well as presentations, publications, and invited keynote addresses.

While flourishing professionally, his personal life had taken a toll. Relocating from his home country to the United States, Mike lost many social ties and relationships. Still, he enjoyed the experience and the social environment of graduate school. While working on his dissertation, he started dating and got married shortly after finishing graduate school. Mike and his wife, born in the United States, decided to move to Europe when he was offered a job at a high-performance training center. Soon after relocating, Mike found that a perceived loss of social ties and family support from moving to yet another country – and this time as an entire family – while at the same time trying to establish himself professionally in the high-performance environment proved more difficult than expected.

Things got even more challenging when his first, and later, second child arrived. It seemed impossible for Mike to juggle the multitude of demands, and to balance time with his wife and kids with professional responsibilities. Lack of self-care became a more pressing issue in coping with the multiple demands of consulting, teaching and research. Traveling for sporting events and conferences, working weekends at matches, tournaments or competitions, engaging in supervision and continuing education as well as recertification requirements, all meant spending time away from his loved ones. Still his strong work ethic, "pushing through" mindset, and perfectionism validated and reinforced his behavior, at the cost of disregarding self-care. He normalized chronic stress symptoms and considered struggles as part of the job he knew he was good at; a job that meant a great deal to him and where he was making a difference to his clients, teams and organizations, students, future professionals, and the wider sport psychology community.

Unfortunately, it was not only the long hours, working weekends and nights, and many roles and responsibilities that were taking a toll on Mike. While he was trying to be there for his wife who needed his help and as the father of their kids and as a husband, while adjusting to living in a new country as a family, Mike experienced increasing difficulties at work. He realized that politics, administration, and value-conflicts started to interfere with his work. His values and person-centered approach were at the core of his professional practice. The system he was working in, however, did not hold the same values and approach. Several conflicts arose where

Mike stood true to his values and acted as an athlete advocate. As a result, the relationship with his boss and colleagues suffered, who eventually bullied him more and more frequently. His everyday life at work – the job he loved and was passionate about – started to cause additional stress. Mood changes, deep exhaustion, irritability, and racing thoughts were symptoms he experienced rather frequently. With pressures and demands coming at him from all sides of his personal and professional life, he who was an expert in burnout, was now struggling with increasing symptoms of chronic stress and burnout himself.

Unexpectedly, Mike found help and support from trusted friends and colleagues in his international network of sport psychology professionals. They invited him to talk about his experience at an international conference. The group held a symposium where they shared details about their journey, accompanied by some of their closest colleagues. Mike talked about what caused his burnout, how he experienced it and the lessons he learned. He shared many of his challenges in building an applied sport psychology career and how some of his personality traits and value-driven professional philosophy contributed to chronic stress – both in the professional and personal spheres of life. As he said that day, "It comes down to practicing what you preach. Having been used to tolerating pain, working through it as an athlete, leads right into doing the same as a professional – even though you know better. Still, granting yourself the right to engage in self-care, making it a priority even ... it's hard."

He shared how he slowly started to realize that something wasn't right. Some of the early symptoms involved insomnia, deep exhaustion, racing thoughts – especially at bedtime, which Mike was shrugging off. But symptoms started to get worse; he started to experience panic attacks and became more and more irritable; not so much at work, but mostly with his kids and wife. He started to withdraw from social engagements, which included spending time with his wife. She would share her worries, for she was definitely seeing his symptoms, and in response he was becoming more and more distant and going into even stronger denial about having any issues. As more and more physical symptoms started to show, it became harder for Mike to deny what was happening to him. Skin rashes, greying hair, and back pain were some of the new symptoms. Eventually, he experienced shortness of breath and blood circulation problems that kept him from exercising and playing ball with his kids, which had been vital activities in helping him recover, decompress, and find perspective. At the same time, feelings of worry, guilt and shame haunted him. How could he resolve this situation and still provide for his family? Wasn't all this his fault anyway? Maybe he'd failed as a professional? He was ruminating over these thoughts.

The tipping point, leading him to finally face the truth and seek help, came when his wife eventually set an ultimatum; either he'd go and seek professional help, or she'd have to leave him. Facing this immanent loss in his personal life, Mike saw a doctor, who diagnosed him with burnout and put him on sick leave. He received professional help from a counselor to support him in working through his personal and professional burnout experience. Eventually, he also committed to resolving the issues he had been ignoring over the years related to his wife, family, and marriage.

Being on sick leave meant that Mike was able to focus on his recovery and through counseling therapy and mindfulness exercises, he was able to gradually overcome his acute burnout episode. He decided to cut down on some of his professional engagements, prioritizing his consulting work. Learning to say 'No' wasn't easy, yet very important for sustaining a life-less riddled with chronic stress. He also formed important alliances that helped, and continue to help, him manage work politics. "Ultimately, it comes down to the 'Three Ds': Define – Decline – Delegate. You have to learn to say no, and prioritizing work and assignments facilitates delegating certain tasks. The most important and challenging thing, however, is (re-)defining your role, your mission, your values, your path. It's easy to lose sight of that, being focused in a rigid way only on advocating on behalf of your clients – neglecting your own personal needs. That's why self-care is so essential," Mike concluded in his presentation.

Fellow professionals attending the session thanked Mike for sharing his personal experiences so openly. Several acknowledged that they share similar experiences. Reflecting on the session, the topic of self-care, establishing professional boundaries and preventing burnout, seems to be a prevalent issue for sport psychology professionals.

Thought Questions

Case Conceptualization

- In light of Mike's story, his experiences, personality traits, situational factors, symptoms experienced, actions taken to overcome his burnout and lessons learned from the experience, what are key aspects of a case conceptualization for sport psychology professionals regarding prioritizing self-care and preventing burnout?
- What key aspects of Mike's past are essential to understanding his burnout experience?
- What external and internal factors might be playing a part in him struggling with chronic stress?
- Professionally and personally, what is important to consider as Mike is navigating transitions, trying to meet expectations and fulfill his multiple roles?

Ethical Considerations

- You are one of Mike's close friends and colleagues and he shares his struggles with you. Considering the symptoms Mike describes, where do you set boundaries as a professional and friend in directly supporting him? How would you tackle a potential resistance to seek professional help?
- How would you like to be treated by close friends and colleagues if you ended up in a similar situation?
- What potential "red flags" exist in Mike's story that point to him being in need of help from a therapist? What kind of intervention would be appropriate and at what stage?

Diversity Awareness/Reflection

- What contextual knowledge of sport (psychology) culture do you currently have that may guide and assist you in this highly demanding and highly competitive setting? What contextual pieces are missing that you would like to know more about?
- How might the fact that Mike and his wife are from two different countries, while living in a third, impact the building of his professional career and the settling of his family?

Intervention Planning and Evaluation

- For professionals like Mike who are experiencing burnout, what steps should they take to seek help? Which potential complementary clinical and nonclinical aspects should be considered?
- In general, what are key elements of self-care in applied sport psychology and how could they be implemented by professionals in the field to prevent burnout?
- What fundamental lessons learned from Mike's story should be considered to set up educational experiences regarding self-care for early career professionals and/or mental training consultants in training?

Section 6

Consulting Competencies

25 The Influence of Spirituality

"Fidget-Spinning and Philippians: Faith Integration and Sport Psychology Consulting with a Christian Youth Athlete"

Trevor J. Egli and Leslee A. Fisher

United States

Josh is an eight-year-old US European American male (Gen Z'er) who is in third grade and in his first season of competitive basketball. He has heterosexual parents who are married and a sister who is two years older. Josh's family is considered upper-middle-class; Josh's father is a full-time physician and Josh's mother works full-time at home. They live in the southeastern United States in a primarily European American neighborhood.

Josh goes to a private evangelical Christian school (Baptist-affiliated) which he has attended his whole life. It is both a day and boarding school, and there are many international students; however, the student population primarily consists of other European Americans. Josh goes home every day and does not board at the school. He is considered one of their best performers in every sport he has played thus far. He excels in soccer (so does his team; they recently went undefeated) and he is a natural runner. However, he does not feel the same pressure playing soccer or running cross-country that he feels when playing basketball. His soccer team is sponsored by a local church and it feels more relaxed and recreational to him, and he primarily runs cross-country to be with his friends. Even though he enjoys participating in both soccer and cross-country through the school, basketball is his favorite sport. There are high expectations from both his parents and coaches for him to perform well in basketball this year due to his athletic ability, basketball skill set, and his natural love of the sport.

During the first couple of games of the season, Josh has fun and performs well. His team is undefeated. He is playing the way others are expecting him to play. In the next game, however, Josh's team was matched against a team that was very "intense" and more aggressive than he had played against before. The players were much bigger than him as well as very physical and emotional (e.g., one player received a technical foul because of his aggressive behavior). On top of this, both the opposing coach and fans were screaming at Josh's team the entire game. As a result, and during the game, Josh "shut down."

DOI: 10.4324/9781003089629-31

Since then, Josh has become fearful of teams with bigger players and reputations for poor sportspersonship, and more specifically, those displaying aggressive behaviors. Josh reported not wanting to play against these teams; however, he knew he would have to due to the limited number of tournaments and teams in the area.

Although this is the first time Josh's team has played against these types of athletes, it is not Josh's first experience with aggressive players. Outside of basketball, Josh still enjoys playing sports of all kinds at home and at school. He often plays with one of his best friends who lives in his neighborhood; however, this friend was also considered a bully by both Josh and his mom. They were both in the same grade in school, loved sport, and lived in the same neighborhood which led to a natural connection; however, Josh is a very gentle, quiet kid, but his friend is much more aggressive in all types of play. Ultimately, he would dominate Josh in whatever activity in which they were involved. This primarily occurred in Josh's neighborhood or at school. His friend bullied Josh by teasing him, "trash talking" when playing sports, and being rude when they were playing together. Josh's typical response was to "just do something else" or go somewhere else even though he wanted to do the original activity he and his friend were doing. Despite feeling bullied at times by his friend when playing sports, he still would participate in these activities with him because he was so interested in playing sports.

During the season, Josh's parents noticed a distinct difference in him when it came to playing in all of his games. In practice, he was excited to play and gave his full effort; however, in games, he was extremely nervous and anxious. In fact, he got physically sick the night before the games and did not want to play. His parents still brought him to the game, but on several occasions, Josh refused to either start the game or return to the game after coming out for a substitution. Coach Johnson always started Josh and he still wants him to start. Both his parents and coach believe that by having him sit out, the team's chances of winning decrease. As a result, Josh felt a lot of pressure to do well when he is in the starting lineup.

In summary, Josh's parents had become very concerned with their son's drastic attitude change and competitive and somatic anxiety surrounding basketball. They tried to encourage him to continue to play and have fun. They employed several different methods to try to help him become less nervous such as bribing him with candy, having more hangouts with friends, and allowing him to not attend games if he did not want to; however, at other times, they forced him to go and "face his fears." Unfortunately, their strategies did not create the kind of change they were hoping to see in Josh. Therefore, they decided to seek out a certified mental performance consultant (CMPC) – you – to help Josh decrease his anxiety and increase his confidence for the remaining month of the season. After doing an internet search, they found the list of CMPCs in their geographic area from the Association for Applied Sport Psychology's website (AASP; www.appliedsportpsych.org).

Josh's father then reaches out to you by e-mail with a brief description of his son's change in behavior during the basketball season. You call Josh's father after receiving the e-mail to get more information on the specific services they are hoping to receive for their son and describe your performance consulting approach. He also asks if you integrate evangelical Christianity into your consulting since you work at a Christian university. Your performance consulting philosophy does include faith integration (when appropriate), which often includes incorporating Christian disciplines, such as prayer and reflection on Bible verses. After agreeing that your services may be helpful, you set up a time to meet with Josh. You send his parents an informed consent form (which includes parental release and child assent) and an intake form (i.e., demographic information, experience with performance consulting, list of potential topics, and reason for seeking out services) that you would like Josh to complete with his parents and bring to his first session with you.

Josh's mother brings him to his first meeting with you. When they arrive at your office, you invite them both in and collect their completed paperwork. The completed intake form contains a request to learn more about competitive anxiety, self-confidence, and integrating faith into sport, all of which were items you discussed over the phone with Josh's father. Of note, there is also mention of Josh having issues with one of his teammates, which included a name next to it. After completing introductions, you ask Josh about what his parents have discussed with him in regard to meeting with you in order to clarify what he can expect. You then ask Josh and his mother if it would be fine with them if you meet with just Josh from now on, and they both agree. Then, Josh's mother waits just outside your office during every session and comes back in at the very end of each session.

Thought Questions

Case Conceptualization

- What information do you need prior to beginning to work with Josh in order to enhance your effectiveness?
- What do you believe are the primary factors that are contributing to Josh's competition anxiety and lack of confidence?
- What developmental factors should you consider when working with someone Josh's age?

Ethical Considerations

- Do the facts in this case suggest any "red flags?" That is, do you notice the need to consider clinical referral, limits of your consulting competency, competing stakeholders, etc.?
- Josh is a minor. What type of relationship will you have with Josh? His parents? What role(s) will you and they play?

- Is it appropriate to integrate evangelical Christian faith into your consulting, as requested by the parents? If so, to what degree? How might you discuss this with Josh? His parents?

Diversity Awareness/Reflection

- What are your initial thoughts regarding the culture(s) involved in this case? These may be related to the performance arena or the individual performer.
- How comfortable are you in working with an individual who ascribes to evangelical Christianity? How familiar are you with this faith tradition?
- Do you believe there should have been a more intensive family inventory taken related to their Christian faith? For example, you do not know how often they attend church, Josh's level of church involvement, or if he participates in other Christian-focused activities. Would having this additional information change your consulting approach?
- What cultural factors may play a role in Josh's performance anxiety and lack of self-confidence in basketball? How do you believe his cultural background will influence his experience of faith?

Intervention Planning and Evaluation

- Identify the potential issues that may be affecting the performer (these may be personal and/or performance-related). From your list, prioritize the top three issues that you believe are having the most immediate impact.
- How might you address competitive anxiety with a child?
- If you do decide to integrate faith into your consulting with Josh, how might you integrate evangelical Christianity into your sessions together?
- You only have one month to work with Josh. Is it possible to be effective in such a short time? If so, what areas do you believe would be most effective to address with Josh?

Additional Questions

- This case study focuses on evangelical Christianity. How comfortable would you be working with an individual who ascribes to a different religion and/or spirituality?
- If you are uncomfortable and/or unfamiliar with a particular religion and/or spirituality of an individual, what steps might you be able to take to become more culturally competent and to better help the individual?
- Do you feel it would be appropriate to incorporate parents into the performance consulting relationship when working with a minor?

26 Working with Coaches

"Consulting Sports Coaches: A Case of Coach Development"

Alexander T. Latinjak

SPAIN

The Coach

Paul was a 34-year-old tennis coach working at a large western European tennis academy when we started working one-on-one. We had known each other for a number of years as I had worked with several of his players providing my sport psychology services. When he reached out to me to work with him personally, I had been working as an external consultant to the academy and my job included working individually with players and staff who asked for help. Hence, I was paid by the academy and not by Paul himself.

Since Paul and I had been talking as colleagues for years about players he coached, I thought it would be essential to differentiate the new kind of work he asked that we do and chose a structured interview format for our first one-on-one sessions. That way I intended to establish clear new roles for our commencing collaboration. In these structured interviews, we first talked about his reasons to approach me and his expectancies from our collaboration. Later, we explored relevant external factors related to his work, and he described himself globally, in the sport context, and in specific situations. I also asked him to reflect on his psychological skills, searching for strengths, weaknesses, and specific coping strategies. In the following sections, I summarize the results of this initial exploration.

Aims and Expectancies

Paul's main reason for seeking to engage with sports psychology on a personal level was a sense of demotivation and dissatisfaction with his professional life. He felt trapped in a job that offered few prospects and that he could not do forever because of the physical demands of working eight hours a day outdoors all year round. For example, he told me that one day he felt so devastated because he had to work on the court all day that he locked himself in his car for more than half an hour while his players warmed up alone. He hoped that our collaboration would help him cope better with his current job and develop viable plans to further develop his career as a tennis coach.

DOI: 10.4324/9781003089629-32

His dream was to coach professional players, and at that time he could not see a way to reach his goal. He wanted to travel with professional players to tournaments and guide them in their path to world elite. He imagined being part of the culture of top tennis and interacting with the sport's elite. Paul himself never was a professional player and he admitted having little idea what it actually takes to be a top coach. He read many books written by elite coaches and tennis players, yet he felt astray, and without a clear idea what steps to follow to reach his aims. He often complained that there was no established route or professional education to become a world-class tennis coach. From our work together he hoped to acquire the confidence needed to be able to cope with the demands of coaching on the top level.

External Variables: The Academy He Worked At

Paul worked at a large tennis academy with over 100 players, some of whom were among the best in their age group nationally and even internationally. On paper, the academy had all the facilities it needed to promote future top tennis players. From the courts to the gym, and even the weather, were great for working with junior tennis players. For Paul, however, it was how the club functioned internally that had become a major struggle.

During his more than 6 years at the academy, he was responsible for groups of players between the ages of 12 and 14. As soon as the players improved sufficiently, they moved to more advanced groups and started working with other coaches. According to Paul, the players had long forgotten about him by the time they reached the age they wanted to leave the academy and maybe take one of their coaches with them. Also, according to Paul, the players believed that the better a coach, the higher the group he or she trains in the academy. Since he coached one of the groups of younger players, he believed that he was less well perceived as a coach than others, by players, colleagues, and the academy management.

In addition, when players are good enough to play professional tournaments, they usually turn to coaches who have already worked with elite players before. Of course, that was not him. Paul described professional coaching as a fairly closed society to which very few, and definitely not he, had access. Overall, he summarized the external variables in his job as a system in which he would never get promoted, forcing him day in and day out to do a job that was too physically demanding with low financial reward and low future prospects.

Psychological Descriptors

Paul described himself as an ambitious loner. Too strange to be close to family or have too many friends; too driven to be content with what he was. He wanted to, dreamed of, traveling with players all over the world, maybe to compensate that he was never even close to making it as an elite tennis player himself. In a clash with this ambition of his, he felt that the perceived realities of him having no idea how to make his dreams come true, nor having the confidence to believe that it would actually ever happen, were countering any ambition.

He said that in the morning it was getting harder and harder to get out of bed to work with players, half of whom were there just because their parents wanted to park them somewhere. The other half of his players would eventually leave him anyway once he did a good job making them better. In the evenings he felt exhausted, sunburned in summer and with back pain in winter.

Every now and then he felt inspired and threw himself into work. Soon after, however, he would realize that all his efforts would not help his own development. Once he realized that the academy managers did not value his work, or at least not make it perceptible to him, he would feel even more frustrated and less satisfied with his work and life. He would have loved to quit his job with a public dramatic act if possible, but he had no other training and was not interested in the few jobs (e.g., waiting tables) that were available to him.

Psychological Skills

Paul mentioned many psychological strengths, including the ability to spot technical flaws in players, make himself understood to players, connect with players off the court, work hard, focus solely on the job, overcome setbacks, and to invest in his future at the expense of the present. However, Paul also recognized several key weaknesses that hampered his professional development. First, he never managed to develop a winner's attitude. In his role as a coach he lacked confidence during competitions and felt like he could not pass a positive attitude on to his players. He also lacked the social skills to build a good network with other coaches, so he never managed to make friends with someone who actually coached professional players.

Dealing with the situation was also far from ideal. For some time, he had resorted to all sorts of medication because of his multiple pains and sleep problems. At times when he felt particularly depressed, he felt the need to have a few drinks in the evening to make up for a long day at work. He had even tried some recreational drugs over the weekend to experience emotions and connect with other people. He did not express acute concerns about these behaviors, although he recognized that they could become problematic if his situation did not change soon.

At the End of the Exploratory Interviews

After completing the interviews, I worked with Paul on a visual representation of the resulting key concepts. Paul assessed each concept based on its valence and relevance. Paul also took a copy of the figure home as a takeaway message. The graphical representation became a basis of our collaboration over the next six years. We made changes to it by adding concepts and changing the ratings of valence and relevance as needed.

Thought Questions

Case Conceptualization

- Which are the key external variables, psychological descriptors, and psychological skills (including coping strategies) that explain Paul's situation and that can be key to solve some of his issues?
- Which external variables, psychological descriptors, and psychological skills (including coping strategies) should rather be considered consequences of key variables, thus somewhat distracting from the main issues?

Ethical Considerations

- Considering that the academy is initially paying for the consulting service, what limitations or boundaries does that imply for working with Paul?
- How would you deal with potential reasons for referral based on clinical (burnout) or health (drug consumption) disorders?

Diversity Awareness/Reflection

- How does the context of professional tennis differ to that of other sports, and how does that impact the case?
- How does the western European cultural context influence Paul's case, and in what degree would the case be different in other cultural contexts?
    - Which parameters regarding the case, the context and the person would you consider differently and how?
- Paul is a male tennis coach; in what way(s) would him being a female tennis coach working in a western European cultural context would change the case?
    - Which parameters regarding the case, the context and the person would you consider differently and how?

Intervention Planning and Evaluation

- After reviewing the key variables that explain Paul's case, argue about the time frame you would set for the collaboration, the options being days, months, a few years, or many years?

- Argue to what degree the work with Paul should be directed to (a) changing external variables which affect Paul, or (b) teaching Paul skills for self-control so that he positively affects/changes his environment?
- Define a set of realistic short-, mid-, and/or long-term goals and measures that you could use at the start and end of the collaboration to ascertain the degree of success or effectiveness of your work. Think how success could be defined in Paul's case and whether there are one or more ways in which the collaboration can be successful.

27 Working with Parents

"It Runs in the Family"

Kathy Feinstein

United States

Mr. and Mrs. Ramirez are the proud parents of two daughters: Leanna, 12, and Lizette, 9. Much of their pride centers around the girls' accomplishments in school, in sports, and in being good citizens. Mr. and Mrs. Ramirez reflect similar values in their current lives and family histories. Neither are strangers to hard work. They both immigrated from Mexico during childhood and watched as their parents made their way in a new country through hard work and family support.

Mr. Ramirez knew as a young boy that he wanted to play professional soccer. Upon moving to the United States, he continued to pursue his dream and worked hard to manage his training, schooling, and job to meet his parent's expectations. Through dedication, Mr. Ramirez became an accomplished soccer player. In fact, he ultimately reached his goal of playing soccer for his home country on a semi-professional team. He remembers this as the best time of his life. It was short-lived, however, when his career came to an end after suffering a knee injury. He was devastated. However, he was also in love with the love of his life, a welcomed distraction that helped soothe his sorrow. Mr. and Mrs. Ramirez were married shortly thereafter. Although he never competed at the same level, Mr. Ramirez continued to play soccer until his family and work obligations did not provide the free time that he needed to be a reliable teammate.

On the other hand, Mrs. Ramirez was always a sports fan but never played sports growing up. As a young Latina, she was encouraged to take on more traditionally feminine interests and hobbies. She enjoyed her studies and was also active in her church.

Not surprisingly, the Ramirez couple took on traditional roles in their marriage. Now Mr. Ramirez runs the family-owned landscaping business. At the same time, Mrs. Ramirez manages the bookkeeping and caring for their two daughters and the couple's nieces and nephews when their siblings call.

The Ramirezes provide a comfortable lifestyle for their family, including gymnastic team training and private coaching for both daughters. Traveling the state for competitions requires the family to allocate significant resources to support their daughters. Specifically, the family commits

DOI: 10.4324/9781003089629-33

considerable financial resources for travel and training as well as a substantial amount of time for traveling on weekends during competition season. In return, their expectations are that the girls work hard at their sport while maintaining maximum effort in school, do their chores at home, and actively participate in volunteer experiences for their church.

Both parents hope that sports participation will help their daughters develop self-confidence, a good work ethic, and learn how to be valuable teammates. Further, they believe that as long as their daughters are doing their best by giving their best effort, the results don't matter. Mr. Ramirez believes his success today is directly related to the attributes he developed playing sports, and he has the same dream for his girls.

Although the sisters have consistently experienced success in gymnastics, they approach the sport differently. Lizette, the younger sister, approaches gymnastics aggressively, and her parents describe her as "fearless." Whereas Leanna takes a more thoughtful and systematic approach. Subsequently, Lizette is racing through the levels. In fact, Mrs. Ramirez worries that she will outperform her older sister, which she fears will devastate Leanna.

Leanna, a straight-A student, showed promise in gymnastics early on, excelling from the time she began training approximately five years ago. After two years of training, she quickly advanced through the levels and began competing on an Xcel team. Social gratification bolsters Leanna's passion and love for the sport. Her best friends were her gymnastics teammates, and she looked forward to practices and competitions. Two years prior, during an Xcel Silver competition, she experienced a "mental block" and could not complete a back handspring in her floor routine during the competition. She was not sure what happened and, at first, decided it was an isolated incident and continued her training and competing. However, she continued to struggle and began feeling "scared" but could not articulate what she was afraid of.

Now Leanna cannot move onto a level 5 team without proficiency in a series of movements, including a round-off back handspring back tuck. Her coach has given her a deadline, and she must have her backward moving movements and be ready to compete by the end of the training season.

Her parents have been supportive and patient. They enlisted the help of another coach, her favorite from seasons past, who took her back to basics. Subsequently, over the coming months, there were several changes in the coaching staff at the gym. Mr. and Mrs. Ramirez viewed this as an opportunity for Leanna to see her block through several different lenses. However, the changes in coaches seemed to create more stress for Leanna. She feels pressure to perform and worries her coaches will be "upset" and view her as weak if she does not get her back handspring back.

As time has gone on without much progress, Mr. Ramirez has begun feeling impatient. After all, he played semi-professional soccer in Mexico and remembered feeling stressed and nervous before a competition, too, but sitting on the bench was never an option. Mr. Ramirez can't understand why Leanna cannot just do the back handspring like he's seen her do countless times in the past. In his desperation, he tells her he will not pay for more lessons if she is not going to "try" harder to overcome her adversity. Further, in his attempt to motivate her, he compares her to her sister three years her junior and asks why she can't do what she does?

As a student-athlete, Leanna not only takes her gymnastics seriously, but she also prioritizes her studies. She begins to channel her efforts at school. However, recently Leanna lamented about having a "B" average in one of her classes. She contacts her teacher as soon as her grade drops. She is desperate to know how to improve her grade before the end of the quarter. She cannot tolerate feeling like a failure in both her sport and school.

Leanna enjoys the big sister role and takes pride in "coaching" her little sister in gymnastics. She often sets the air track mat up in the backyard so they can practice together. Recently she has noticed how little fear her sister seems to have as she practices, and Leanna remembers when she felt like that. Leanna is proud of her sister and, at the same time, thinks about what it would be like to have her sister make a level 6 team while she is left behind. Leanna keeps these conflicted feelings to herself because she does not want to upset anyone in the family.

Leanna's struggle with the mental block consumes the whole family. Mr. and Mrs. Ramirez see the solution differently. Mrs. Ramirez prefers to listen to the coaches and give Leanna space and time to work through the block. Mr. Ramirez wants results. Understandably this causes considerable stress between them, and they both feel quite helpless in their quest to help Leanna overcome her mental block. As a gymnast, Leanna cannot move onto a level 5 team without being proficient in a series of movements, including a round-off back handspring back tuck. With the looming deadline, the Ramirez family is concerned that Leanna will not succeed without professional help and support. At this point, Mrs. Ramirez calls you to help her daughter "get" her movements back. You share your initial consultation information, including fees, and Mrs. Ramirez agrees to consult with her husband and call back to schedule an appointment. Mrs. Ramirez calls again one week later to schedule the appointment.

Thought Questions

Case Conceptualization

- What are some ways that Mr. Ramirez's experience as a semi-professional athlete may be affecting the way he is parenting his athlete?
- What family strengths exist that support Leanna's goals?
- Leanna's sister's success creates conflicting feelings for Leanna, which may impact her performance. How might either or both parents influence Leanna's emotional response to her sister's progress while she works to overcome her mental block?

Ethical Considerations

- Leanna is 12 years old. How will you define the terms of confidentiality?
- Leanna is the client; however, her parent's involvement is crucial to the treatment. How will you define boundaries?
- What contextual knowledge do you have about mental blocks in gymnastics?
- What contextual knowledge you have about family systems/sports parenting?

Diversity Awareness/Reflection

- What contextual knowledge do you have of the LatinX culture that might influence your work with this family?

Intervention Planning and Evaluation

- What interventions can the consultant use to effectively mitigate the gap between Mr. Ramirez's understanding of Leanna's mental block and his reality that a mental block is not a real phenomenon?
- Parents typically bring their athletes to consultants to improve their performance. How will you include/invite the parents into understanding their role in supporting their athlete overcome a mental block in your intervention planning?

Index

Made in the USA
Monee, IL
13 September 2026

62978976R00094